SCIENCE AND SYNTHESIS

An International Colloquium organized by Unesco on the Tenth Anniversary of the Death of Albert Einstein and Teilhard de Chardin

---

RENÉ MAHEU

FERDINAND GONSETH

J. ROBERT OPPENHEIMER

WERNER HEISENBERG

REVEREND DOMINIQUE DUBARLE

SIR JULIAN HUXLEY

GIORGIO DE SANTILLANA

GERALD HOLTON

B. M. KEDROV

FRANÇOIS LE LIONNAIS

LOUIS DE BROGLIE

RENÉ POIRIER

PIERRE AUGER

JEAN PIVETEAU

REVEREND PIERR

French Contributions Translated by

BARBARA M. CROOK

SPRINGER-VERLAG BERLIN · HEIDELBERG · NEW YORK 1971

Translated from the French by Barbara M. Crook, with the exception of the papers by J. R. Oppenheimer, W. Heisenberg, Sir Julian Huxley, G. Holton and B. M. Kedrov, also certain contributions to the debates, which were originally delivered in English.

Original publication in French by Gallimard.

Softcover reprint of the hardcover 1st edition 1967
Library of Congress Catalog Card Number 77-143044

ISBN-13: 978-3-642-95208-1 e-ISBN-13: 978-3-642-95206-7
DOI: 10.1007/ 978-3-642-95206-7

SCIENCE AND SYNTHESIS

Contents

Introduction ix

Speech of Welcome *by René Maheu* xi

PART ONE

Albert Einstein and the Scientific Synthesis

Speakers

F. Gonseth: *Einstein's Knowledge of Nature and Philosophy* 3

Werner Heisenberg: *The Unified Field Theory* 12

Reverend Dominique Dubarle: *Science and the Unified Vision of the Universe. Einstein's Ideas and Teilhard de Chardin's Contribution* 17

Sir Julian Huxley: *Science and Synthesis* 28

G. de Santillana: *The Great Cosmological Doctrines* 37

G. Holton: *Where is Reality? The Answers of Einstein* 45

B. M. Kedrov: *Integration and Differentiation in the Modern Sciences. General Evolution of Scientific Knowledge* 70

PART TWO
Science and Synthesis: Debates

---

1. FROM PLURALITY TO UNITY

Chairman	F. le Lionnais	77
Main Speaker	Louis de Broglie	78
Others	Reverend F. Russo	84
	J. Ullmo	86
	A. Lichnerowicz	88
	J.-P. Vigier	91
	A. Matveyev	93
	F. Gonseth	99
	V. Kourganoff	100
	O. Costa de Beauregard	101

2. TOWARDS A COSMOLOGY

Chairman	F. le Lionnais	
Main Speaker	R. Poirier	103
Others	A. Lichnerowicz	109
	O. Costa de Beauregard	112
	J. Merleau-Ponty	114
	G. Cocconi	117
	Stamatia Mavridès	118
	V. Kourganoff	120
	A. Trautman	124
	W. Heisenberg	125
	B. d'Espagnat	126
	J.-P. Vigier	128
	J. Ullmo	129

3. DETERMINISM AND INDETERMINISM

Chairman	F. le Lionnais	
Main Speaker	W. Heisenberg	134
Others	O. Costa de Beauregard	135
	A. Matveyev	138
	J. Ullmo	140
	B. d'Espagnat	142
	J.-P. Vigier	143
	Reverend D. Dubarle	145
	J.-L. Destouches	146

4. THE ORGANIZATION OF SCIENTIFIC RESEARCH

Chairman	P. Auger	147
Speakers	B. M. Kedrov	151
	P. Piganiol	154
	G. Holton	156
	M. Debeauvais	161
	V. Kourganoff	165
	Reverend P. Leroy	166
	Reverend F. Russo	169
	F. le Lionnais	173

PART THREE

Teilhard de Chardin

Main Speakers	J. Piveteau	179
	Reverend P. Leroy	184
Others	O. Costa de Beauregard	186
	H. de Terra	191
	P. Chouard	192
	Madeleine Barthélémy-Madaule	194
	F. Meyer	198

Index 203

Introduction

This volume is a collection of the lectures and discussions at an international colloquium organized by Unesco on the theme of *Science and Synthesis* to mark the occasion of the 10th anniversary of the death of both Einstein and Teilhard de Chardin, also the 50th anniversary of the theory of general relativity.

Despite the great gulf which lies between the work of Einstein and Teilhard de Chardin, the coincidence in the dates provided an opportunity to examine the urge towards a synthesis of the scientific and philosophical approaches which lies at the very heart of the work of these two great men. It was, indeed, their common desire for an all-embracing concept of the universe which led them both to try to construct a cosmology for the modern world.

So it seemed that the best way of honoring Einstein and Teilhard de Chardin was to arrange a free discussion of the current likelihood of effecting a synthesis of scientific knowledge which would bring together some of today's most eminent scholars, inspired by the wish to make their research more meaningful by philosophic reflection. The resulting give and take of ideas would go far beyond mere commemoration; it would bring their ideas to life by setting them against the present state of science.

This book falls into three parts. The first, which bears the name of Einstein, takes its cue from his work and goes on to discuss synthesis in the physical sciences; it comprises an eye-witness account by Professor Gonseth of Einstein's earliest discoveries at Zurich and Berne: *Einstein's knowledge of nature and philosophy as revealed in his work;* a paper by Professor J. Robert Oppenheimer on the man who was his colleague at Princeton; a communication by Professor Werner Heisenberg: *Einstein and Synthesis — the Unified Field Theory,* which pursues the dialogue on determinism and indeterminism which had gone on for years between the two scientists. The contributions of the Reverend Father Dubarle, Academician B. Kedrov and Sir Julian Huxley extend the debate from pure physics to the life sciences and other scientific disciplines; finally, Professor Giorgio de Santillana locates the modern cosmologies within the historical perspective of the ancient cosmological doctrines.

The second part consists of Round Table debates, chaired by Mr. François le Lionnais, on three themes: *Einstein — from Plurality to Unity; Towards a Cosmology;* and *Determinism and Indeterminism.* The first was introduced by the Duc de Broglie, the second by Mr. René Poirier and the last by Professor Heisenberg. This part also includes the debates chaired by Professor Pierre Auger on synthesis as a method of organizing modern science (coordination of scientific disciplines and research planning at the national and international level).

The third part records the exchanges at another Round Table debate on the work of Teilhard de Chardin, the theme being: *Knowledge of Nature and Man.* Mr. Costa de Beauregard, the Reverend Father Leroy and Professor J. Piveteau took part in these discussions.

So it will be seen that this colloquium on *Science and Synthesis,* taking as its starting point the physical sciences, proceeded via the problems of synthesis in the life sciences to those of the human sciences.

The opinions expressed by the various authors and the points of view adopted by them are entirely their own and do not necessarily represent those of Unesco.

Speech of Welcome

BY RENÉ MAHEU
Director General of Unesco

Your Excellencies, Ladies and Gentlemen:

It was Easter Day, 10th April, 1955. Father Teilhard de Chardin lay dying in New York; he was 74. Yet at the end of a long life, alternately and sometimes even simultaneously brilliant and seclusive, he found only two loyal friends to follow his coffin to the grave. A week later, on 18th April, 1955, not many miles away at Princeton, Albert Einstein, his senior by a bare two years, passed away at the height of his fame, leaving the entire world with a sense of irreparable loss which persists to this day.

The historian of the future may perhaps pause to reflect upon this dramatic coincidence. He may contrast the difference in the destinies of these two great and solitary men of genius, utterly different in their work, their research and their beliefs, yet spanning the same historical period and suffering its wars and persecutions, and he may see in the conjunction of their passing one of the really significant events of this mixed-up age we live in.

We are met together, 10 years after the death of these two great men, to honor their memory and bear witness to the perennial freshness of their thoughts and ideas. This is the primary objective of this Colloquium in the course of which the almost legendary personalities of Einstein and Teilhard will be freely discussed within the context of contemporary science. You will hear Professor Gonseth and Professor J. Robert Oppenheimer, both of whom knew Einstein well, talking about him as a man. This meeting, which is unconventional in many ways, is, however, more than a simple act of commemoration. Moreover, no homage could be more welcome to those we seek to honor than to keep their memory green by relating problems to which they devoted their lives to today's living reality.

It would be wrong to suppose that the aim of this Colloquium is a confrontation of their ideas and systems. Such a confrontation, however interpreted, would presuppose the existence of some basis of comparison, if not actual assimilation, and this is emphatically not the case.

We are not here to consider their respective niches in the history of science. We are here for quite another purpose: to stimulate debate on a series of questions which, as they knew better than anyone, transcend man's greatest efforts and constantly require examining afresh. If in the course of discussion their ideas are quoted in support of the argument, it will be only as major illustrations, highly significant, of course, but in no way exclusive.

Our general theme is *Science and Synthesis*. The aim of this Colloquium is to set the procedure and conquests of science — that is to say its methods and findings — against the demands of intellectual synthesis, as by definition required for our concepts of man and the universe. I need not tell you that the work of Einstein and Teilhard is full of typical examples of such endeavors. The mathematical model of the physical laws of matter which the former created in his theory of general relativity and the interpretation of the evolution of the forms and dimensions of life which the latter developed from palaeontology and eschatology are unquestionably, each in its own way and at its own valuation, the vastest and at the same time the most intensive systems of knowledge ever conceived. But it is not enough to say that never before had there been such an ambitious attempt at synthesis in the scientific field; we must add that never before had synthesis been so consciously and deliberately identified with the true essence of science as in the minds of these great men.

Nor need I explain at length the supreme importance Unesco attaches to questions of the kind inspired by the example of Einstein and Teilhard on the theme of Science and Synthesis. It has been said before and it will be said again, Unesco is an organization with a humanistic vocation. Everything it undertakes implies a particular view of man, and it tries to promote and realize this aim step by step but on a universal scale. Everything it does, technical and specialized though it may appear, has a single aim, a single direction and a single dimension, engaging the whole man in each of us and the whole of mankind which is also in each of us. Thus, Unesco is by its very nature dedicated to the spirit of synthesis, and we must remain vigilant to see that no temptation towards misplaced erudition, no urge towards greater efficiency, drags it so far into specialization that it forgets the vocation which inspires its ethical mission.

On the contrary, in face of the growing specialization of thought and action brought about by diversification in research and the division of labor, Unesco has a duty to promote interdisciplinary activities and contacts and to encourage broad views, in short, to emphasize the vital importance of the spirit of synthesis for the health of our civilization. I say vital advisedly since man — and I mean his essence, which is to say his judgement and his freedom of choice — is just as likely to be smothered by his knowledge as paralysed by the lack of it. Similarly, he is quite as likely to lose his identity in the confusion of competing social pressures as to atrophy in the condition known as under-developed.

This is why a Colloquium such as this is of the greatest value to Unesco by reason of its purpose, and not only by reason of the distinction of those taking part in it.

When studying the program, you may have noticed that we have tried not to impose a rigid framework which might inhibit the spontaneity of the evidence and the communications, or restrict the freedom of the declarations and discussions. An occasion such as this would be highly vulnerable to any sort of dictatorial approach. However, I think it is in order and may perhaps help to clarify the general idea of the enterprise if I indicate in advance the various planes and dimensions within which the lectures and debates are arranged. As I see it, there are three of these planes or dimensions, corresponding to the three aspects of science and synthesis: first, the study and appreciation of the content and structure of the synthesis of knowledge insofar as it is an accomplished fact; second, the examination of the nature and scientific value of synthesizing thought regarded as agent and intellectual design; and finally, the determination of the conditions under which synthesis may be achieved within the framework of the organization of scientific work.

How far have we progressed with the effective synthesis of what we know? Or, to put it in another way, how far have we progressed with our knowledge of the universe and mankind, with particular reference to the two great syntheses of Einstein and Teilhard? Further, what do we think today of the scientific standing of these two great men? These questions form the main theme of the Duc de Broglie's introduction to the Round Table debate on *Albert Einstein and the Co-existence of Waves and Particles,* as they do, too, of Professor Santillana's talk on *The Great Cosmological Doctrines* and Professor Holton's on *Where is Reality? The Answers of Einstein,* Professor Heisenberg's on *Einstein and Synthesis: the Unified Field Theory* and Professor Piveteau's on *Teilhard de Chardin and the Problem of Evolution.* The same spirit will inform the debates on *Knowledge of Nature and Man,* relating to Teilhard's work, and those on the theme, *Towards a Cosmology: Determinism and Indeterminism.*

A further perspective is the one I have already mentioned, the study of the nature and epistemological value of creative thought and universal synthesis. The synthesizing process is, of course, essential to the maintenance of balance among all the parts and at all the levels of science. This will be explained by the Reverend Father Dubarle and Sir Julian Huxley in their general considerations and subsequently discussed by Professor Kedrov in his paper on *Specialization and Integration in the Modern Sciences.* I question whether the partial and total integrations spring from ideas of the same sort and of equal worth, and, even supposing that they do, there is still another question which must be asked: are all branches of science and all instruments of scientific thought equally capable of achieving unity? Louis de Broglie will give us his views on this when he discusses *The Reduction of Plurality to Unity in Einstein's Work,* and so, too, will Professor Gonseth on *Knowledge of Nature and Philosophy in the Work of Einstein.* "Knowledge of philosophy" — this is the key expression. How do we know how much of the total synthesis of knowledge is science and how much philosophy? And again, under what conditions can an attempt at total synthesis claim to be both scientifically rigorous and philosophically significant? The answers may well be scattered over the whole of the distance which separates Euclid's *Theorems* and Archimedes' *Principles* from the myths of Plato. Which is to say that this is the sort of question which calls forth serious differences. It is, however, a question to be asked, indeed, a question to be pressed, and the right people to ask are the distinguished scholars and thinkers whom we are honored to see here today. Put to minds of this quality, this question has a meaning which is clearly of vital importance to all of us. Let us turn it over, as one might turn over a coin; we then perceive that what it means is this: is there a valid *single total* thought object? And, more particularly, nearer home, I might say, what is in the face of this insistence on total Oneness the reality of Man, who is at one and the same time a part of the natural order and the measure of the universe, a stage in evolution and the conscience of history? Is he an object, a process or a limit? Or again, is this demand for Oneness a challenge from without and, if so, whence? Or is it a self-imposed demand, as inseparable from man as the meaning from the object, and which is his abstract aspect?

As you see, the answer or answers concern us all, and concern us most intimately. And precisely because the answers science gives to these key questions concern the whole of mankind, we should remember that science does not aspire merely to achieve a unity of knowledge but also, following the profound ideas of Auguste Comte, a unity of minds. Science, before it becomes an encyclopedia, is a laboratory, a team and — it is no exaggeration to say this — a society. This society is constantly expanding towards a limit which also tends to be total and which is mankind itself. Therefore, the answers science gives should not be delivered and received as though they were the oracles of an elite, set apart from the rest of human kind, but as the culmination of the work of a community which is growing and becoming more open all the time, a work in which all may and, indeed, should share, if only by an effort at understanding. On this condition alone can science become for everyone what it should be, and what it is for those who live by it: a culture.

The word "society" implies complexity and organization. The problem of synthesis in science does not arise solely at the level of knowledge and thought;

it also arises at the level where scientific work is organized on a world-wide basis. It is true that up to the present the great scientific syntheses, like the philosophical ones to which they are related or to which they lead, have been almost exclusively the work of individual men of genius who towered like mountain peaks above their background and their time. But it is quite possible that the diversification of disciplines and the proliferation of research centers have so changed the conditions for scientific work, even at this exalted level, that theories like those devised by Einstein and Teilhard may well be the last constructive interpretations of man and the universe to emerge from flights of thought as penetratingly individual as those of the poet or artist. Though it is true to say that the organization of scientific research is not of itself productive of synthesis, it is certainly a prior condition and, so to speak, foreshadows it or projects it on to the plane of the society of minds. Into this third dimension of the Colloquium falls the last Round Table debate, to be chaired by Professor Auger.

These, ladies and gentlemen, are the aims as Unesco sees them of the Colloquium to which I have the honor to welcome you. We are indebted to the distinguished scientists who have accepted our invitation to take part in it. The mere fact of their coming together is a significant event. To these scientists and all who have come to hear them and will carry the living echo of their thoughts back to the diverse horizons of their respective countries, disciplines and generations, as represented and assembled here in a single intellectual fraternity, I offer the freedom of our house.

It is my desire that, in addition to the great practical tasks, indispensable though they are, of international co-operation in the spheres of education, science and culture, which are debated and settled in this very room, Unesco's General Conference Room, this building should also provide a home for the universal dialogue on the search for truth and all the doubts and uncertainties, fears and hopes of the human spirit. I desire this, not only to enable them to become better acquainted, but to help them to make their mark on history and to exert a more effective influence on the fate of mankind.

Is this not, indeed, the last message of Einstein and Teilhard de Chardin? Fully aware as they were, in wonder and in anguish, of mankind's splendid and terrible freedom, its ability either to conquer the universe or destroy itself, they both proclaimed the necessity for organizing a world-wide commonwealth, not merely as a condition of progress but as a condition of the very survival of mankind. To this end, both called for the establishment of permanent institutions to facilitate and encourage confrontation and understanding between disciplines, cultures and peoples. These institutions already exist — Unesco is one of them — and it is vital that they should carry out their mission with the support of those for whose benefit they were created. I repeat, it is vital, both for the dignity of mankind and the peace of the world.

There are many pages I could quote from the works of these two great men, but I shall simply end this talk by reading a passage from a letter Einstein wrote to Freud at the time when attempts were being made to establish international co-operation within the framework of the League of Nations: "Today the intellectual elite has no direct influence upon the history of nations; it is too scattered to be of direct help in solving current problems. But this international community could exert an important and salutary moral influence upon the management of politics."

PART ONE

Einstein and the Scientific Synthesis

The Scientific Synthesis

EINSTEIN'S KNOWLEDGE OF NATURE AND PHILOSOPHY

Ferdinand Gonseth Over 50 years ago I, who am speaking to you today, attended the first course Einstein gave after he was appointed Professor of Theoretical Physics at the Zurich Polytechnic. Incidentally, Einstein was to break off this course after a few terms in order to accept an offer from the Kaiser-Wilhelm-Institut, Berlin. This first course was about relativity, of the type later to be called special relativity. At that time the theory of general relativity, i. e. Einstein's theory of gravitation, had not yet been formulated. It was, however, beginning to take shape and during the last few months of Einstein's time at Zurich rumors of something strange and mysterious were going around. I diligently attended his other courses, too, and I can still vividly recall a course in mechanics which managed to be both strictly classical and highly original. Along with my fellow-students in the Department of Mathematics and Physics, at that time very few in number, I also attended his seminar at which we were privileged to talk with him — though I fear we were unaware how great a privilege this was!

Before he went to America, Einstein returned several times to Zurich to expound the successive versions of his theory of gravitation, the variations being chiefly concerned with the definition of space-time relationships.

As regards the subsequent course of his thinking, and in particular his constantly renewed attempts to define and fix a unified field, that is to say, one which would account for all the physical phenomena, I — like everyone else — had to rely upon his published work.

Another source of information, less direct and yet more intimate, was to open up for me during the 10 years I taught mathematics at the University of Berne. I became friendly with Michele Besso, Einstein's former colleague at the Swiss Patents Office, who had been his closest friend and most trusted confidant. Einstein called him "the best sounding-board in Europe". Besso and I used to take long walks together during which we discussed everything under the sun, but mainly science and philosophy. And so it was I learned of other walks when, with Einstein expounding and Besso contradicting, the theory of relativity underwent and survived its earliest trials.

Did Einstein's personality and authority make an immediate impression on me in 1912 when I was a student of mathematics and physics? Did the originality, the profundity and the truth of his ideas penetrate my mind with irresistible force? A mind too immature and too impatient to acquire knowledge seldom possesses standards by which it can accurately assess the worth of the ideas and personalities it encounters. My mind had difficulty in assimilating Einstein's teaching. The difficulty for me — as for many others — was not so much in following it step by step but in understanding where it came from, where it stood and where it was leading us. Clearly his mind had its own compass, but not only was this hidden from me, it was almost as if something prevented me from seeing it. I have since come to understand that if I had thought I understood him then, I should merely have been deceiving myself.

We have come a long way since then. True, there are still some unsolved problems on the physicist's horizon, while other even more intractable problems have arisen since. But along the horizon of method a new light has dawned; of course, the illumination is still not perfect, but what right have we to expect that it should be? Today the methodological compass which so surely guided Einstein's thought is public property. There have even been some improvements made in it, as happens with all useful scientific instruments. To realize what it meant at that time, we have only to look — with the eyes of today, from the position assured by the success we currently enjoy — at the position Einstein occupied then and introduce a little systematization into it. It is an honor to do so on the present occasion, but it is no longer an adventure.

In his first course on special relativity, to which I have already referred, Einstein never so much as mentioned a four-dimensional universe (in which time is taken as the fourth dimension), or used terms different from those of Minkowski's space-time relationships. On other occasions he demonstrated his complete mastery of these. Having recourse to space-time can mask certain bold assumptions; Einstein's way of going to work, which incidentally is now well known, threw them into relief. Contrary to what appeared obvious, he postulated that the speed of light would be the same for all observers moving at constant speed, whatever that speed. Going on from there, he carefully considered how all the clocks in the same system, which would inevitably be linked with the same observer, could be synchronized by light signals alone. He then established how one could relate in a coherent manner the times and distances measured by two observers in a state of constant speed (i. e. in a state of inertia) by relating them to one another. This way of going on had devastating consequences. The principle of simultaneity, for example, was lost, but Lorentz' formulas were rediscovered and with them — and without the need to postulate an ether possessing spectacular properties — the explanation of phenomena then crying out for elucidation, such as the negative result of the Michelson-Morley experiment, the Doppler effect and so on. Well, more than 50 years afterwards, I can still recall the feeling of insecurity and even worry this type of reasoning produced in me. Is it not a somewhat arbitrary act to postulate that the speed of light is constant *before* synchronizing the clocks? And can its arbitrariness be expunged by the value of its consequences? Does the gain in efficiency make up for the loss in evidence? How was he to make us understand that these questions, which many other people have put to themselves, go nowhere near the heart of the problem!

I began to gain some insight into the matter as I listened to Einstein himself expounding the dilemma which confronted physicists at that time. There was, he said, a difficult choice before us: we had either to discard classical kinematics and the "obvious" formula for the composition of speeds, or jettison Maxwell's equations. For Maxwell's equations are not invariant, as it would seem that they ought to be for the transformations which enable the system of reference to be changed in classical kinematics. (To be precise, these transformations are the analogs in classical kinematics of the Lorentz transformations in special relativity.) However, Einstein added, there was no real freedom of choice for the physicist — it was unthinkable that we should abandon Maxwell's equations.

What Einstein mentioned in very summary fashion and with never a thought of analysing it, was not the validity of this or that train of reasoning, the result of this or that measurement, or the value of this or that interpretation, but rather a whole web of reasoning, results and successful applications at the center of which Maxwell's equations assumed a special significance. Now, the solidity of such a web and the security it gives to a scientist can be so compelling that, when it comes into conflict with some of the most firmly anchored proofs, it is opting for the proofs which lays him open to the risk of very grave error. In such a situation, theoretical reason is no longer sole arbiter; but neither can the practical reasons be on their own. Here is a complex of circumstances and consequences to be appreciated and evaluated. What must be perceived is what is best retained or discarded, taken or left. This can be made into a principle, that of the greatest suitability, which I prefer to call the principle of the best fit. But to observe or apply such a principle is not in the least like applying a regular sequence of algorithmic or automatic procedures. The use of this principle requires a mind formed by its possessor's own experience and informed by that of others. Minds with native judgement exist, as a sort of natural phenomenon, but even for them the choice of the best fit carries some risk; their judgement, their choice, their decision can never be guided by the verdict of experience since it has not yet been acquired.

As regards the option from which the theory of relativity sprang, experience has woven around it such a web of confirmation and success that doubt is simply driven out. But this is something everyone knows, and it is not what I wish to demonstrate here and now. What I wish to stress particularly is the methodological aspect of Einstein's initiative. With unequalled simplicity and naturalness, he assumed something which now seems essential equipment for the research scientist: that he should be aware of both his freedom and his responsibility. He must insist upon complete freedom of investigation, yet know, too, that this freedom has a built-in hazard, that of arbitrary affirmation. He must at the same time be open to the evidence of the facts, yet know that this openness, too, has a built-in hazard, that of being deceived by appearances. The freedom and the compliance could well be in conflict; they are not fixed in advance. The scientist remains the arbiter, the principle on which he bases his decision being always the best fit — and of that, he can be the sole judge.

It may surprise you that this is what I consider the crucial clause in the research scientist's statute book; you may object that it is always present and always has been in true scientific research. Granted, it has never been absent, but it has been crushed under the weightiest and most diverse hypotheses. Albert Einstein set it free from this burden. Never before had freedom of investigation gone so far as to make the proofs give way to it, nor had the throwing open to experience conceived its field of action in such bold terms. And yet, never before had a procedure for gaining knowledge proved so incisive and effective. Never before, in a word, had a method been such a good fit.

Now, research which opts simultaneously for freedom of investigation and openness to experience and undertakes to reconcile them with a view to the best fit, has by this very act made itself methodologically and philosophically autonomous. It is in a position to reject any philosophy which is not its own, any philosophy whose principles antedate it or are foreign to it. Let us say, rather, taking the argument to its logical conclusion, that research which adopts this method accepts responsibility for its central philosophy, that of knowing best, of knowing the full range of what is possible. Now, as far as knowledge of nature is concerned, no philosophy ever went this far. It is clear that, when it does not dwell upon the particular, it is the most faithful realization of philosophical intention. I have said that to be fair and see it straight, we have to look at Einstein's method with the eyes of today. This is exactly what I have been doing. Did I not give my findings an unduly determinist bias, and did I not in laying so much stress on Einstein's early work leave aside other data which might have led to different conclusions? In order to have a more or less complete picture, we should, of course, have to analyze the method which enabled the theory of special relativity to unfold into a generalized relativity; we should also have to put it in its proper place and interpret the intention Einstein never abandoned, of giving form and reality to the unified field; and finally, we should have to account for his refusal to rally to basic indeterminism. The portrait of Einstein which I have assembled after much thought would not be modified thereby. The essential features were present right from the start. Albert Einstein's philosophical autonomy was already apparent in an answer he gave at his first lecture course in Paris to someone who asked him what he thought about Kant, or what he made of him. "Ah," said he, "Kant is all things to all men."

All in all, I think I may sum up what I have said in a few words: in Albert Einstein, the scientist was the natural philosopher incarnate. As a philosopher he was profound, though he might perhaps have been considered naive if he had not been so clear-sighted. I might say that in him the scientist was nothing less than the perfected form of the philosopher, set free by his sincerity and the genuineness of his search.

I have assembled a number of ideas suggested and supported by the case of Albert Einstein; but they go beyond the particular case. I must therefore explain what guarantees of their correctness are offered by the situation in which we find ourselves at present. To do this, I need give only a few brief indications, and in doing so I shall refer to the goal of systematization, of which I have already spoken, and explain its significance at the same time.

As I have said, research which opts simultaneously for freedom of investigation and openness to experience and undertakes to reconcile them with a view to the best fit, has by this very act made itself methodologically and philosophically autonomous. And, as I have explained, once the attitude and the option had been adopted, scientific research could do no other than conform, whether consciously or not, to a greater or lesser degree.

Let us consider what makes it into a methodology. In the practice, in the ups and downs of research, one can make out the conditions without which it could not flourish; one can also make out the consequences which are inseparable from it. And so one sees, not without a certain dismay, a coherent methodology emerging, the open methodology, the proofs of whose rightness are to be found in the success of the procedure itself.

Is this methodology applicable only to research concerned with natural phenomena? As regards the study of man, there are now many indications that the procedure — with the addition of openness to introspection and the use of controls — is well on its way to becoming applicable there.

As to philosophy in general, the principles of best fit and openness can also be applied here. I have every reason to think and believe that they are capable of illuminating the truth of that famous saying: "Science and philosophy form a single whole".

To return for the last time to my subject, I believe that the seeds of all that followed were contained within his assumption of the freedom to discard a proof in favor of a conviction born of the pursuit of truth.

EINSTEIN'S PRESENCE*

J. Robert Oppenheimer † It is an honor to be in this company, the tenth year after Einstein's death, the fiftieth anniversary of his discovery of the general theory of relativity.

As the President has said, though I knew Einstein for two or three decades, it was only in the last decade of his life that we were close colleagues and something of friends. But I thought that it might be useful, because I am sure that it is not too soon — and for our generation perhaps almost too late — to start to dispel the clouds of myth and to see the great mountain peak that these clouds hide. As always, the myth has its charms; but the truth is far more beautiful.

Late in his life, in connection with his despair over weapons and wars, Einstein said that if he had to live it over again he would be a plumber. This was a balance of seriousness and jest that no one should now attempt to disturb. Believe me, he had no idea of what it was to be a plumber; least of all in the United States, where we have a joke that the typical behavior of this specialist is that he never brings his tools to the scence of the crisis. Einstein brought his tools to his crises; Einstein was a physicist, a natural philosopher, the greatest of our time.

What we have heard, what you all know, what is the true part of the myth, is his extraordinary originality. The discovery of quanta would surely have come one way or another, but he discovered them. Deep understanding of what it means that no signal could travel faster than light would surely have come; the formal equations were already known; but this simple, brilliant understanding of the physics could well have been slow in coming, and blurred, had he not done it for us. The general theory of relativity which, even today, is not well proved experimentally, no one but he would have done for a long, long time. It is in fact only in the last decade, the last years, that one has seen how a pedestrian and hard-working physicist, or many of them, might reach that theory and understand this singular union of geometry and gravitation, and we can do even that today only because some of the *a priori* open possibilities are limited by the confirmation of Einstein's discovery that light would be deflected by gravity.

Yet there is another side besides the originality. Einstein brought to the work of originality deep elements of tradition. It is only possible to discover in part how he came by it, by following his reading, his friendships, the meager record that we have. But of these deep-seated elements of tradition — I will not try to enumerate them all; I do not know them all — at least three were indispensable and stayed with him.

* Title added after the author's death.

The first is from the rather beautiful but recondite part of physics that is the explanation of the laws of thermodynamics in terms of the mechanics of large numbers of particles, statistical mechanics. This was with Einstein all the time. It was what enabled him from Planck's discovery of the law of black body radiation to conclude that light was not only waves but particles; particles with an energy proportional to their frequency and momentum determined by their wave-number, the famous relations that de Broglie was to extend to all matter, to electrons first and then clearly to all matter.

It was this statistical tradition that led Einstein to the laws governing the emission and absorption of light by atomic systems. It was this that enabled him to see the connection between de Broglie's waves and the statistics of light-quanta proposed by Bose. It was this that kept him an active proponent and discoverer of the new phenomena of quantum physics up to 1925.

The second and equally deep strand — and here I think we do know where it came from — was his total love of the idea of a field: the following of physical phenomena in minute and infinitely subdividable detail in space and in time. This gave him his first great drama of trying to see how Maxwell's equations could be true. They were the first field equations of physics; they are still true today with only very minor and well-understood modifications. It is this tradition which made him know that there had to be a field theory of gravitation, long before the clues to that theory were securely in his hand.

The third tradition was less one of physics than of philosophy. It is a form of the principle of sufficient reason. It was Einstein who asked what do we mean, what can we measure, what elements in physics are conventional? He insisted that those elements that were conventional could have no part in the real predictions of physics. This also had roots: for one the mathematical invention of Riemann, who saw how very limited the geometry of the Greeks had been, how unreasonably limited. But in a more important sense, it followed from the long tradition of European philosophy, you may say, starting with Descartes — if you wish, you can start it in the thirteenth century, because in fact it did start then — and leading through the British empiricists, and very clearly formulated, though probably without influence in Europe, by Charles Peirce: one had to ask how do we do it, what do we mean, is this just something that we use to help ourselves in calculating, or is it something that we can actually study in nature by physical means. For the point here is that the laws of nature not only describe the results of observations, but the laws of nature delimit the scope of observations. That was the point of Einstein's understanding of the limiting character of the velocity of light; it also was the nature of the resolution in quantum theory, where the quantum of action, Planck's constant, was recognized as limiting the fineness of the transaction between the system studied and the machinery used to study it, limiting this fineness in a form of atomicity quite different from and much more radical than any that the Greeks had imagined or that was familiar from the atomic theory of chemistry.

In the last years of Einstein's life, the last twenty-five years, his tradition in a certain sense failed him. They were the years he spent at Princeton and this, though a source of sorrow, should not be concealed. He had a right to that failure. He spent those years, first, in trying to prove that the quantum theory had inconsistencies in it. No one could have been more ingenious in thinking up unexpected and clever examples, but it turned out that the inconsistencies were not there; and often their resolution could be found in earlier work of Einstein himself. When that did not work, after repeated efforts, Einstein had simply to say that he did not like the theory. He did not like the elements of indeterminacy. He did not like abandonment of continuity or of causality. These were things that he had grown up with, saved by him, and enormously enlarged; and to see them lost, even though he had put the dagger in the hand of their assassin by his own work, was very hard on him. He fought with Bohr in a noble and furious way, and he fought with the theory which he had fathered but which he hated. That is not the first time it has happened in science.

He also worked with a very ambitious program, to combine the understanding of electricity and gravitation in such a way as to explain what he regarded as the semblance — the illusion — of discreteness, of particles in nature. I think that it was clear then, and I believe it to be obviously clear today, that the things that this theory worked with were too meager, left out too much that was known to physicists but had not been known much in Einstein's student days. Thus it looked like a hopelessly limited and historically rather accidentally conditioned approach. Although Einstein commanded the affection, or, more rightly, the love of everyone for his determination to see through his program, he lost most contact with the profession of physics, because there were things that had been learned which came too late in life for him to concern himself with them.

Einstein was indeed one of the friendliest of men. I had the impression that he was also, in an important sense, alone. Many very great men are lonely; yet I had the impression that although he was a deep and loyal friend, the stronger human affections played a not very deep or very central part in his life taken as a whole. He had of course incredibly many disciples, in the sense of people who, reading his work or hearing it taught by him, learned from him and had a new view of physics, of the philosophy of physics, of the nature of the world that we live in. But he did not have, in the technical jargon, a school. He did not have very many students who were his concern as apprentices and disciples. And there was an element of the lone worker in him, in sharp contrast to the teams we see today, and in sharp contrast to the highly co-operative way in which some other parts of science have developed. In later years, he had people working with him. They were typically called assistants and they had a wonderful life. Just being with him was wonderful. His secretary had a wonderful life. The sense of grandeur never left him for a minute, nor his sense of humor. The assistants did one thing which he lacked in his young days. His early papers are paralyzingly beautiful, but there are many errata. Later there were none. I had the impression that, along with its miseries, his fame gave him some pleasures, not only the human pleasure of meeting people but the extreme pleasure of music played not only with Elizabeth of Belgium but more with Adolphe Busch, for he was not that good a violinist. He loved the sea and he loved sailing and was always grateful for a ship. I remember walking home with him on his seventy-first birthday. He said, "You know, when it's once been given to a man to do something sensible, afterward life is a little strange."

Einstein is also, and I think rightly, known as a man of very great goodwill and humanity. Indeed, if I had to think of a single word for his attitude towards human problems, I would pick the Sanscrit word "Ahinsa", not to hurt, harmlessness. He had a deep distrust of power; he did not have that convenient and natural converse with statesmen and men of power that was quite appropriate to Rutherford and to Bohr, perhaps the two physicists of this century who most nearly rivalled him in eminence. In 1915, as he made the general theory of relativity, Europe was tearing itself to pieces and half losing its past. He was always a pacifist. Only as the Nazis came into power in Germany did he have some doubts, as his famous and rather deep exchange of letters with Freud showed, and began to understand with melancholy and without true acceptance that, in addition to understanding, man sometimes has a duty to act.

After what you have heard, I need not say how luminous was his intelligence. He was almost wholly without sophistication and wholly without worldliness. I think that in England people would have said that he did not have much "background," and in America that he lacked "education." This may throw some light on how these words are used. I think that this simplicity, this lack of clutter and this lack of cant, had a lot to do with his preservation throughout of a certain pure, rather Spinoza-like, philosophical monism, which of course is hard to maintain if you have been "educated" and have a "background." There was always with him a wonderful purity at once childlike and profoundly stubborn.

Einstein is often blamed or praised or credited with those miserable bombs. This is not in my opinion true. The special theory of relativity might not have been beautiful without Einstein; but it would have been a tool for physicists, and by 1932 the experimental evidence for the inter-convertibility of matter and energy, which he had predicted, was overwhelming. The feasibility of doing anything with this in such a massive way was not clear until seven years later, and then almost by accident. This was not what Einstein really was after. His part was that of creating an intellectual revolution, and discovering more than any scientist of our time how profound were the errors made by men before then. He did write a letter to Roosevelt about atomic energy. I think this was in part his agony at the evil of the Nazis, in part not wanting to harm any one in any way; but I ought to report that that letter had very little effect, and that Einstein himself is really not answerable for all that came later. I believe he so understood it himself.

His was a voice raised with very great weight against violence and cruelty wherever he saw them and, after the war, he spoke with deep emotion and I believe with great weight about the supreme violence of these atomic weapons. He said at once with great simplicity: now we must make a world government. It was very forthright, it was very abrupt, it was no doubt "uneducated," no doubt without "background;" still all of us in some thoughtful measure must recognize that he was right.

Without power, without calculation, with none of the profoundly political humor that characterized Gandhi, he nevertheless did move the political world. In almost the last act of his life, he joined with Lord Russell in suggesting that men of science get together and see if they could not understand one another and avert the disaster which he foresaw from the arms race. The so-called Pugwash movement which has a longer name now, was the direct result of this appeal. I know it to be true that it had an essential part to play in the Treaty of Moscow, the limited test-ban treaty, which is a tentative, but to me very precious, declaration that reason might still prevail.

In his last years, as I knew him, Einstein was a twentieth century Ecclesiastes, saying with unrelenting and indomitable cheerfulness, "Vanity of vanities, all is vanity."

THE UNIFIED FIELD THEORY

Werner Heisenberg Among the many ideas which Einstein has pursued in connection with his theory of general relativity, his proposal of a unified theory has aroused the widest interest on account of its philosophical implication. Einstein has suggested that such different phenomena as gravitation, electromagnetism and material bodies could ultimately be described by one fundamental field or system of fields; that all the different empirical laws of nature could be expressed by one universal system of non-linear equations for the components of this field. From a philosophical point of view this possibility looks very attractive. Different groups of phenomena, like gravitation and electricity, can scarcely be separated completely. They may influence each other, and therefore the laws of nature responsible for them cannot be completely independent. The unified field theory would contain the different laws as special cases and would at the same time establish the connection and thereby state the structure of nature.

Einstein was not able to carry this program very far. His starting point was the field of gravitation for which the field equations were given by the theory of general relativity. He then intended to find a field structure which would be a natural generalization of the symmetrical (metrical) tensor, representing gravitation, as well as a system of field equations for this structure which would represent a natural generalization of the equations of pure gravitation. In a first attempt he tried to include the electromagnetic laws; with regard to the material bodies, he hoped that at a later stage of the theory the elementary particles could be understood as singularities in space of the universal field. This hope was motivated by the non-linear character of the field equations which might lead to such singularities. But at this point he ignored — one may almost say, by intention — the quantum theoretical nature of the elementary particles and therefore he could not possibly find a correct mathematical description of their behavior.

Before going into the details of this question we have to mention another important problem, the connection between the system of field equations and the cosmological model of the world. Einstein saw this connection in the light of the ideas proposed by Mach. The rotation of a single body in empty space has, according to Mach, no meaning. Therefore a centrifugal force can occur only if space is not empty, if distant masses produce this force. Hence the reaction of a single body on its motion depends on the distribution of matter in the universe. This distribution and the corresponding structure of space-time is not uniquely determined by the field equations. But it is not completely arbitrary; it is limited by the field equations and should correspond to one of the many solutions of the field equations. The behavior of a single particle under the influence of local fields may then to some extent depend on the structure of the universe. It is true that Mach's principle is not so intimately connected with Einstein's field equations as Einstein had believed. But the relation between the cosmological model of the world and the field equations, the relevance of this cosmological structure for the behavior even of small bodies remains an essential feature of any unified field theory.

Coming back to the quantum theoretical nature of elementary particles, we first notice that singularities in space produced by a classical non-linear field equation would behave quite differently from real elementary particles in a given field of force. All those features, which in quantum theory are connected with the apparent dualism between the pictures presented by waves and particles and which are expressed by the mathematical scheme of quantum or wave mechanics, would not be seen in the behavior of the singularities. Therefore in our time it would be a quite unrealistic approach to try to connect different groups of phenomena in nature without taking quantum theory into account from the very beginning.

Besides that, the many experiments carried out with the help of the big accelerators during recent years have given us a great wealth of information about elementary particles, not accessible for Einstein in his time. We have learned that besides electromagnetic forces and the corresponding photons, besides gravitation and the corresponding gravitons, there exist very many different fields of force, each characterized by the corresponding elementary particle, for example: those forces which bind an atomic nucleus together. A unified field theory would have to comprise all those different fields. When two elementary particles collide at very high energy, many new particles emerge from the collision; we speak of multiple production of particles. But such phenomena would not be well described by saying that the particles have been broken into many smaller pieces. It is much more correct to state that the big kinetic energy of the colliding particles has been transmuted into matter — following Einstein's law — by the creation of many new elementary particles. Actually, whatever the special nature of the colliding particles may have been, the emerging particles always belong to the same well-known spectrum of elementary particles. Energy becomes matter by assuming the form of an elementary particle. The spectrum of elementary particles reproduces itself in the high-energy collision processes.

A number of very important conclusions can be drawn from these results. One can see at once that any attempt to construct a separate theory for each of the hundred different fields of force would be absurd. The unified theory may have been an object of speculation for Einstein; in our time it is an absolute necessity in theoretical physics if we want to understand the elementary particles.

One may perhaps doubt whether the future theory will be a unified field theory, or whether other mathematical tools than fields could be more adequate for the description of the experiments. But it must be a unified theory comprising all the different empirical fields.

Einstein had believed that the particles were singularities of the field in space. In quantum field theory we have learned in the meantime that the particles are singularities — namely poles — in momentum space, not in ordinary space. For Einstein the field was real, it was in fact the ultimate reality and determined both the geometry of the world and the structure of the material bodies. In quantum theory the field distinguishes, as in classical physics, between something and nothing; but its essential function is to change the state of the world, which is characterized by a probability amplitude, by a statement concerning potentialities. In this way experimental situations in elementary particle physics can be described by applying operators constructed from products of field operators on the groundstate "world". But one can scarcely consider the fields as real and objective in the same sense as Einstein did in his field theory.

Both in Einstein's theory and in modern quantum field theory, the final formulation of the underlying natural law is given by the field equation. Therefore the central problem of the unified field theory is the correct choice of the field equation and the comparison of the results with the experimental observations. In this respect any attempt at a unified quantum field theory is in a much better position than Einstein's older theory. So many details are known nowadays about the spectrum of elementary particles, their interactions, selection rules in transitions etc., that it should be comparatively easy, in spite of the great mathematical difficulties, to see whether a special field equation suggested as fundamental law has a chance to give results in agreement with the many observations.

If one tries to find the fundamental field equation as a result of an analysis of the experiments, the most important information is obtained from the laws of conservation, selection rules and empirical quantum numbers. Already forty years ago the physicists had learned from the mathematicians that these relations are due to symmetries, to "group properties" in the underlying natural law. Therefore the empirical information will reveal the group structure of the fundamental field equation, and it may well be that the group structure — perhaps together with a few other plausible postulates — determines this equation uniquely.

The analysis of the spectrum and of the selection rules would be a straightforward method for determining the group structure of the underlying natural law, if all observed symmetries were exact symmetries. This, however, is not true: there are approximate symmetries like the isospin group, and higher groups like SU_3, SU_6, SU_{12} etc., which hold only in a very rough approximation. In this case one has no choice but between two possibilities. One may either assume that the underlying law is strictly symmetrical under the group concerned, but that the symmetry is broken later on by an asymmetrical groundstate. Or one may assume that the symmetry is not contained in the underlying law, but that the approximate symmetry is produced indirectly by the dynamics of the system. The two possibilities can be distinguished by an experimental criterion. In the first case one should, according to a theorem of Goldstone, observe bosons (particles obeying Bose statistics) of rest mass zero, responsible for breaking the symmetry. In the second case such particles should not exist. For the isospin group, one actually observes the electromagnetic field and the photons of rest mass zero which are responsible for the violation of the symmetry. For the higher groups, SU_3, SU_6 etc., such particles have not been seen. If one takes this as the final result of the analysis, one arrives at the conclusion that the underlying natural law should be invariant under the Lorentz group, the isospin group and a few gauge groups (the latter for baryonic, leptonic number, strangeness and electric charge). There is just one simple nonlinear differential equation containing these symmetries, and it is therefore natural as a trial to take this equation as a basis for the unified field theory. The differential character of the equation emphasizes the relation between cause and effect which is sometimes called relativistic causality. Relativistic causality is compatible with the statistical character of quantum theory, and its consequences seem to agree well with the observations of collision processes.

Starting from this non-linear spinor equation one, arrives at a number of encouraging results which in my mind make it probable that this equation is already the correct basis of elementary particle physics. But I cannot go into any details. Instead of discussing special consequences of this unified quantum field theory, I will try to compare its general structure and its results with Einstein's earlier program. The center of the new theory is formed by the strong interactions in which most elementary particles, baryons and mesons, participate and which have the full symmetry of the equation. The strongly interacting particles and the corresponding fields had not been considered by Einstein in his attempts at a unified theory, partly because he could not accept the quantum theoretical relation between fields and particles, and partly because very few of those particles and fields were known in his time. Therefore in this respect the two theories are very different.

The electromagnetic field however was included in Einstein's attempt; it appears in the unified quantum field theory as a rather special kind of field resulting from the asymmetry of the world under the isospin transformations. At this point the new theory has revealed a most interesting relation between the macroscopic structure, the cosmological model of the world and the properties of the elementary particles. This relation has been expressed in a somewhat mathematical form as a theorem by Goldstone. If the underlying natural law is invariant under certain transformations (in this case, the transformations are isospace), and if this symmetry is broken by an asymmetry of the groundstate "world", the theorem states that necessarily bosons of rest mass zero must appear, or — changing over from particles to fields — long-range forces. These forces make it understandable that the properties of the particles cannot be completely independent of the macroscopic structure of the world. Actually, the number of protons in the world is very different from the number of neutrons; therefore the real world is not invariant under rotations in isospace. At the same time, we know that the electromagnetic forces have long range; the corresponding particles, the photons, have rest mass zero. Therefore, it looks very natural to assume that the electromagnetic field, or parts of it, represent a Goldstone field and that its existence is due to the asymmetry of the world in isospace.

This result emphasizes the close similarity between the forces of inertia (for example: centrifugal force) and their cosmological origin in Einstein's theory, on the one hand, and the electromagnetic forces with their cosmological origin in the unified quantum field theory, on the other hand. In both cases, a qualitative assumption about a fundamental asymmetry in the cosmological model is sufficient to determine the forces uniquely and quantitatively. In general relativity the value of the centrifugal force follows when one knows that, at large distances, the metric approaches the Euclidean metric. In quantum field theory, the strength of the electromagnetic field or the elementary charge are determined when one knows that the macroscopic world is asymmetric under rotations in isospace. It is encouraging to see that the value of the electric charge — or, what is equivalent to it, the value of Sommerfeld's fine structure constant — comes out in satisfactory agreement with the observed value, as could be demonstrated in a paper by Duer, Yamamoto and Yamasaki. This result is perhaps the strongost argument in favor of the assumed non-linear field equation.

The field of gravitation was at the center of Einstein's unified field theory. In the unified quantum field theory, gravitation has not yet been considered, and it certainly plays only a very minor role for the spectrum of elementary particles. Still, the general way to the incorporation of the gravitational field seems to be rather clear. It would not be convenient to start, as Einstein had done, with a general Riemannian geometry. Thirring has been able to show in a very important paper, that one may very well start from a field equation invariant under Lorentz transformations, like the non-linear spinor equation. If the fundamental equation leads — among many other asymptotic fields — to a tensor field of long range, then this asymptotic field could have all the properties of the gravitational field. Such a long-range force could again appear in connection with an asymmetry of the groundstate "world", according to Goldstone's theorem. Gravitation would in this way again be a consequence of the macroscopic structure of the world, as in Einstein's theory.

Furthermore, Thirring has pointed out that the behavior of measuring rods and clocks would be influenced by the presence of such a gravitational field. If the four-dimensional geometry in space-time could be measured by real rods and clocks, the result would be a Riemannian geometry of just the type considered by Einstein. Therefore, this geometry is a natural but indirect consequence of the postulate, that the measuring rods and clocks should obey the same universal law expressed by the field equation; that the unified field theory should, as von Weiszäcker has put it, have its inner "semantic", its own consistent scheme of interpretation.

In the present state of physics, we are still very far from a complete solution to all these problems. There are many phenomena in elementary particle physics, and possibly elsewhere, which have not yet been properly understood within the framework of the unified field theory. Still, the program formulated by Einstein's fundamental idea has kept its philosophical force, in spite of, or rather because of, all the new experimental information about elementary particles, and defines in our time perhaps the most fascinating field of research.

SCIENCE AND THE UNIFIED VISION OF THE UNIVERSE. EINSTEIN'S IDEAS AND TEILHARD DE CHARDIN'S CONTRIBUTION

Reverend Dominique Dubarle I believe that this meeting and its theme, and the questions which are inseparable from it, bear most eloquent witness to the intellectual adventure, the culmination of centuries, on which our Western civilization is now embarked. Science can today be seen to be the most typical creation of the thought of this civilization, just as it has become the most indispensable tool of its enterprise. Not many generations ago, science still had to persuade the lay public of its importance. Today this is no longer necessary; daily life is permeated with the ideas of those who crusaded for the consolidation of the practice of science among all sorts and conditions of people. Something most extraordinary would have to happen now to jeopardize the position it has won for itself.

But to say that science has become one of the major intellectual occupations of Western civilization, as well as earning its daily bread, is not sufficient to define with accuracy what is special about our present situation, nor even to give a fair characterization of the role science plays in it. Science is important to us, not only because of the knowledge it brings, or because of what it enables us to accomplish, but also because it is effecting an even more deeply penetrating metamorphosis of the mind. The various components and the multiple lessons of this metamorphosis will be of the greatest moment in the course of future generations for the ideas now unfolding before us and for the humanism of which tomorrow's civilization will have to make a more thorough conquest.

I believe it was this feeling which inspired Unesco when the decision was taken to organize this colloquium. We should be grateful for its discernment in touching upon a vital function of scientific thought, and for its choice of subject for these debates and discussions. The power science possesses to metamorphose our minds is here given pride of place in a form which enables us usefully to discuss its problems, which are far from being solved in our time.

The human mind is spontaneously attracted by unity. Its dream is to be the triumphant creator of this unity, wherever it meets with all life's diversity and the apparent discordancy of things, not to mention the entire range of human conflicts. Of the intellectual excitement of this dream, science was born. For to know with certainty what really happens and to have this recognized as true by the whole world; to say with precision what anybody can understand with precision whilst appreciating that it was well conceived and well expressed; to arrange in due order the multitude of humanity in a single stable community of human intelligence which will increase from one generation to another, this was already in its main outlines the substance of the will to science which the genius of the Greeks shaped so effectively that it survived the breakdown of tradition and language. Today the substance of the modern will to science is twice as strong, after being newly affirmed by the great Seventeenth century initiators, men like Galileo and Descartes and many others. In face of this will, it seems only right and fitting to expect that scientific thought will supply, on the one hand, the unifying, organic and coherent vision of reality as a whole towards which the scientific intelligence always turns and, on the other hand and simultaneously, a unity in the human ways of understanding these things in their wholeness. Under this sign of science, the West has never ceased to hope for the establishment of a firm base for intellectual unity and of a body of knowledge and a principle of synthesizing thought common to all mankind which can be passed on to future generations.

I have described the dream, the will and the hope which inspire the mind set upon following the intellectual path of science; but I dare not declare that what our human intelligence is experiencing at the present time corresponds to the expectations of the dream and the will driving it towards realization. Comparing what has actually happened with what was expected, we have to admit that it is somewhat different. The lesson to be drawn from those events, events intimately concerned with the nature of science and intelligence, during a period which we may suppose to have been crucial for intellectual life, is something we shall have to examine and consider, no doubt from many angles, during the course of this colloquium.

These are not questions put to us from outside, as has often been the case when philosophy has come face to face with science; these questions have grown out of the most advanced scientific thought and are both the result and the function of its present great achievements. They arise from the very works of science, from the thoughts of those who are both originators and privileged witnesses of its accomplishments and sometimes also — and this is a fact I must stress — the prophets who urge us to reconsider certain concepts which are commonly accepted by almost the whole of the scientific community.

Up to the present, Einstein is unquestionably the greatest of the constellation of physical and mathematical geniuses who have since the turn of the century completely upset the ideas taught by conventional science about the universe and the course of natural events. As I shall explain, we owe to him the reopening of a path along which physics can approach a true cosmology, properly scientific, and no longer merely poetic or philosophical, as cosmologies have been in the past. As for Teilhard de Chardin, whose work we are now considering along with Einstein's because of the close coincidence of their demise, we see him here as the animator of the life sciences, the man who pressed into their service his personal gifts of expression and orchestration in order to make clear what these sciences can offer to human thought once it sets out to understand the whole of human experience.

It was certainly a bold stroke to bring together these two systems of thought on the intellectual plane, ignoring all that might be said about the differences in their two brands of humanism and the source of their inspiration. In particular, it may appear to the adherents of physics that the intellectual contribution of the life sciences at their present stage of development is of little value for cosmology and something quite outside the true perspectives of the very fundamental sciences of nature and matter. It is, however, a boldness we should welcome, as helping us to understand the whole of science, including both the life sciences and the more material sciences, at a time when we are embarking on a discussion of the scientific possibility of understanding the whole of reality.

Nevertheless, some order must be maintained in our way of thinking and a distinction must be made between knowledge which is truly scientific and that which its authors have to attribute to sources other than strictly scientific ones. From both these points of view, it is only right that I should give priority to the consideration of Einstein's ideas on the universe and the nature of the wholeness which governs the deployment of these many-faceted realities. For this is, without question, the terrain where the most universally fundamental and the most lucidly scientific ideas come together.

Einstein's ideas now give us a very clear understanding of the extent to which the features of a scientific vision of the universe depend upon the intellectual equipment with which the instrument of scientific discovery is endowed, even more than upon the particular concepts from which it is constructed. When they were first proposed, the ideas underlying the theory of relativity were considered revolutionary. They led to an entirely fresh understanding of both time and the elementary matter of the physical world. Einstein's theory of relativity forces us to realize that the familiar framework of chronological preception, which has borne the seal of scientific respectability ever since Greek astronomy was first established and to which classical mechanics conforms absolutely, is not all that appropriate for describing the total range of physical events in a universe where all forms of matter are in motion relative to one another. Furthermore, this same theory forces us to revise the traditional concept of matter; the primitive idea of the solid and weighty corporality of things now has to give way to the idea of energy as the basic stuff of reality. This means that at the very core of physics the themes and functions of action are displacing the last remaining traces of the scheme of matter. It would be difficult to imagine a more radical revision of the elementary concepts of natural philosophy.

However, once the first intellectual shock was over, once the difficulties of assimilation had been overcome — and even the best minds long bore witness to these difficulties by the naivety of their expositions and intuitive reasoning — it was generally recognized that Einstein's ideas bore the stamp of a great and superior kind of scientific classicism. This was recognized the sooner because, at almost the same time and in step with the studies of the fine structure of things then going on, physics was beginning to promote an entirely new way of looking at elementary nature which is now generally agreed to have undermined the classic canons of science.

Einstein's own attitude helped to reinforce the view which sees in his ideas the last splendid flowering of classical inspiration before the final break occurred. I am speaking, of course, of the ideas which culminated in the theory of general relativity, though we should not forget his 1905 communications on Brownian motion and the photon. The difficulty Einstein himself had in accepting the epistemological revolution implicit in quantum mechanics is well known. He never subscribed to the ideas generally accepted by the theoreticians of this discipline and even today he still provides some sort of justification for those who think these ideas should be re-examined. However, we must admit that he was throughout the leading exponent of the essential propositions of the physical-mechanical theory of nature, renewed and reactivated in all the master works of mathematical physics, from Galileo to Maxwell, via Newton, and later in the analytical mechanics of Lagrange, Laplace and Hamilton-Jacobi.

The physics of relativity is thus characterized by its intellectual fidelity to mathematical analysis and the methods by which it is applied directly to the description of the universe. Its theory is both the extension and the sublimation of the classical tenets of physics and mechanics, as held by Maxwell, Newton and Galileo. A close study of the initiators of these tenets, such as Galileo or, from another point of view, Kepler, produces the astonishing discovery that Einstein's ideas are frequently the fulfilment of the promise of the budding ideas contained within the earlist gropings of modern scientific thought, although the developments which occurred in the intervening time may not have preserved these intentions intact. Thus, Galileo's feeling that motion continues naturally, "by inertia", as we usually express it, is found to be closer to the idea which reappears as the foundation stone of general relativity than the more abstract and classical one which has been accepted since Newton. And again, we find in Kepler's general comparison between the system of Ptolemy, Tycho Brahe and Copernicus considerations which foreshadow in a truly remarkable way the Einsteinian epistemology of the relations between the different descriptions of each individual observer, both between themselves and relative to a common absolute universe. But at the time when modern science was being laid down, the mathematical tools were inadequate for the consistent development of the ideas then beginning to dawn. There was a long road to travel and many powerful resources to be conquered before these ideas could be picked up again. Einstein appeared at a time when science had grown sufficiently mature to enable the rather fleeting intuitive ideas of its first beginnings to be examined afresh.

So, equipped with a more or less adequate mathematical instrument and with a mind formed by the disciplines of pure analysis which were already liberating it from the intangible prison of Euclidian geometry, alerted by the results — still an enigma to his contemporaries — of sophisticated experiments on the behavior of light, Einstein was able to take a firm grasp on physical reality and to handle the phenomena of motion in the way the founders of modern science had dreamed of doing. The epistemology of the observers in special relativity is Kepler's brought to fruition, an epistemology which reveals to the mind something of the invariable of reality which underlies all perceptions. The physics of gravitation in general relativity is the apotheosis of

Galileo's inertia, a physics now able to express its concrete conditions in terms of geometry. The genial artifices of Newton's thought now appear redundant; some of them, indeed, can be dismantled like scaffolding which is needed so long as the monument is under construction, but which could never have been confused with the design of the actual building. And, when this has been done, Einstein's edifice is seen to be admirable because, from its very faithfulness to the inspiration of classical physics and mechanics, it draws a strength which comes from linking the human mind with the universe, something which had been conspicuously lacking in the science of the previous two centuries.

That great historian of scientific thought, Alexandre Koyré, has shown with great force and clarity the cost in conceptual terms, cosmologically speaking, of the advent of modern mechanics. For the human mind, the Copernican revolution was something more than a mere reorganization of the method of representing the solar system. It stood for the intellectual urge to move on to an entirely different way of thinking about the nature of things, as scientifically expressed in Galileo's concept of inertia. At the same time, it supplied an entirely different feeling of oneness. The earlier notion of a well-ordered cosmos, complete in its finite constitution, was to be displaced by the modern idea of a natural universe dispersing all things in the vastness of infinite space, resembling physically the intuitive extension of Euclidian geometry. Man was finding out he did not occupy a privileged position among the stars but was carried hither and thither by the motion of the earth. He was "on the spot" in every sense of the word and in all the physical exposure this implies. And the man on the spot was unable to leave his spot in reality in order to construct a truly universal universe, a scheme of the physical whole. Once the first intoxication with infinity had passed — an intoxication which sparkles engagingly from the naturalistic ideas of Giordano Bruno — and there was time for reflection, our human situation was felt to have something rather unsatisfactory, if not actually disturbing about it. Pascal, most of all, felt this very strongly and incorporated into the modern idea of existence the consciousness of man's disorientation in the vastness of infinite space.

But since the advent of Einstein, the situation so graphically described by Koyré no longer looks so much like the fatal lot drawn by modern scientific thought. Thanks to the magical manipulations of modern differential geometry and a fresh appreciation of the elementary phenomenon of weight, it now seems possible that there may be a viable route leading from the body of knowledge about things acquired by the man on the spot to a universal understanding of the whole of cosmic matter. First, man is able to measure the speed of light and recognize its invariability as inherent in the way each element in the universe is related to all the others. Then, he can think out the facts of gravitation in terms of a new kinematics and discover that this exercise is a good test of the rightness of his ideas: several verifications in good agreement with one another have provided confirmation of the bold concepts of general relativity. Both technically and mathematically there are close links within these concepts between the physics of light and that of gravity. Hence, thanks to the first solid analytical foundation, as well-founded methodologically as the firmest classical concepts, it became possible once again to entertain positive proposals for models of the universe which go beyond the crude illimitation of space and time found in classical physics.

From this point of view, Einstein is the only begetter of scientific cosmology which, before him, was not only non-existent but literally inconceivable for the classical physicists. There is, of course, no question of a return to the cosmos of the ancient Greeks or the mediaeval Schoolmen. But it is a fact that, from this time on, modern science began to glimpse the way back from the classic intellectual projection to infinity of space and time, and to a more unified comprehension of the singularity of diversity. The billions of light years, or the expanding swarm of prodigious numbers of nebuli no longer afflict the spirit of the scientist fortified with the theory of relativity with the feeling of being swamped by infinity. He now possesses the means of integrating these perceptions into structures which are at least scientifically plausible and can draw them together into a single concept of reality. To this concept he can assign a sort of dimension in space and time which, though vast, gives him an idea of the size of the system his mind is endeavoring to encompass.

I think we should pause here to examine this key feature of the Einsteinian vision of the universe before going on to discuss that other vision which, from inside science itself, provides a sort of counterpoint to it. Anyone who has studied, however briefly, the concept of a unitary physical-mechanical universe produced by Einstein's thinking must be struck not only by the inherent satisfaction of drawing all the phenomena together, but also and perhaps even more forcibly by the rhythm of the mathematics in which it is expressed. The exact and unfettered language of almost pure thought, a language whose transparent conciseness at crucial points in the exposition must surprise us, weaves the web of deductive reasoning with an almost magical flexibility. Henceforward, this becomes a natural intellectual necessity, of an inspiration so simple it could be taken for spontaneity or the freedom of a god who, by being what he wills, creates the body he needs.

Thus, we seem to have rediscovered here, fulfilled by science, the inner intention of Spinoza's philosophy which moves from the man on the spot and the dispersed energy of things to the single substance of Nature. Encountering the scholar and philosopher in him, the mind is thus attuned to the supreme intellectual symphony in which are blended together the almost godlike freedom of Nature as a whole and the inescapable necessity governing the behavior of all these fragments. Like Spinoza's philosophy, Einsteinian mathematics, having overcome much of the anguish of the man on the spot and transcended many of the tragedies of human particularity, may be said to introduce us at the last to the intelligent smile of the Universe-God whom the great man greets with a sort of modest brotherliness and of whom he says: "very sophisticated, but not in the least malevolent."

Let me say right away, this is not all of Einstein, nor is it the whole of the great thinker to whom it was granted, as Professor Gonseth has told us, to discover various vital new approaches which opened up fresh avenues in science, nor was it all of the man who often had to fight against the powers of this world, and even against the human implications of that vital piece of the theory of relativity, the equation for converting matter into energy. It was, indeed, during the black year of 1940 that he first introduced this procedure to the powers that be; afterwards, his conscience tormented him most bitterly. But despite all this, I think that what Einstein himself would have put before anything else was the new look the theory of relativity gave to physics, the lesson and perhaps I could even say the poetry of this new look. This man, who in all probability felt this was the unique and exemplary contribution his genius could make to the human search for a unifying understanding, chose to halt on the threshold of a new era of intelligence, insisting on his first model of the scientific vision of oneness. Einstein was, of course, well aware that our perception does not assimilate all that reality offers it. But it was necessary that this first scientific formulation of the great and changing sphere of the generative energy of things, this sphere which perhaps is pulsating for all time like a heart muscle alternately dilating and contracting, should come to provide the basis and the prime element of all subsequent attempts to construct a comprehensive science. Nature is doubtless far more sophisticated than relativity makes her out to be. But our timid intelligence needs reassuring that she is not more impenetrable to ideas nor more hostile than Einstein's vision has shown her to be.

It is time now, however, to take a look at the counterpoint to Einstein's vision of the universe which modern science also began to develop so powerfully some 50 years ago. Any practitioner of physics is familiar with this counterpoint from inside the practice of physics. I evoked it myself a little while ago by the mere mention of quantum physics. However, at the risk of surprising the pure physicist, I shall come at it in a roundabout manner, and I shall begin by asking some questions. The first of these are not connected with physics, but I consider they enable us to make an objective approach, even if it is an oblique one, to reconciling views of the universe which are at first sight so unlike as those of Einstein and Teilhard de Chardin.

Let us leave aside, assuming them resolved, questions about the extreme diversity of the spiritual affiliations of these two scientific personalities, also those manifoldly delicate ones concerning the way in which these affiliations compound with the true and positive contribution of science itself to form the concept of the universe unique to each of them. One fact, however, remains beyond all systematization and criticism: the human experience crystallized in the system of the life sciences proposes to the cosmological idea an intellectual element not directly mentioned in physics. This element in fact operated at the heart of all Teilhard de Chardin's thinking and reappears, variously accentuated, in that of many, if not all, of those who make the living world the theme of their study. Such a fact does not seem destined to remain in obscurity for ever, whether it concerns a vision of the nature of the entire universe which sets out to be as scientific as possible, or whether it concerns physics and the physical foundation of this vision only. Let me explain what I mean.

The positive study of life has revealed the fact of evolution by progressively piecing together the living picture of life. It has uncovered the history of the population of the earth by differentiated species of plants and animals whose particular type is never fixed once and for all, but is the creation of the moment, having a more or less transitory destiny. Over the last 100 years evidence has accumulated to show that this process of becoming also envelops and conditions our human origins, and that the existence of our species is woven into this complex pattern which has its own oneness, formed by the innumerable trajectories of life. This discovery of the origin of life and of evolution has been for many minds a flash of enlightenment, cosmological too, in its way. The best-known works of Teilhard de Chardin are the product of a mind which has experienced this flash with a special intensity.

Now it often happens that, when invoking new evidence for biological evolution, thought — even scientific thought — fails to appreciate the nature of its contribution relative to a synthesis of knowledge and a vision of the whole. Evolution is argued in general terms and, unless one is a physicist, one is blind to the fact that all modern theories of the great systems of reality are also theories of evolution, of new forms of configuration relentlessly thrown up with the passage of time. Ever since classical times, the physical-mechanical explanation has offered a mathematical history of the concatenation of these phenomena. There are very few material systems which science can regard as sufficiently capable of reversible reactions between their component parts, and, moreover, sufficiently protected from external influences, for their processes to occur in perfect cycles. Almost everywhere, and on a universal scale, the distribution of matter and the initial conditions of the processes determine the ordering of the sequence of states of being. In this sense, the universe is evolving physically from the lowest material level upwards.

Thus, it is not merely the simple generality of the idea of evolution which makes the experience of life cosmologically instructive. It is rather the more or less spontaneously perceived fact that with the evolution of life something came into the world that was of a different quality, quite unlike the physical systems and the great astronomical formations. This something has a quality which, first classical physics, and later the physics of the Einsteinian, relativistic image of the universe found they had not the means to homologate and integrate. Like it or not, we must reconcile ourselves to saying this in the language of the non-physicist.

The reality of life in evolution, with its primeval origin, the fact of its existence, its dialectic and its terrestrial Oddysey, looks to our intelligence like the sketch, the draft, the stammering expression, at the same time problematical and convincing, of the *meaning* of that universal reality of which the human species is but one example among many, be it proton, star, metal, flower or fish. And what do I mean by meaning? Applied to the philosophy of life, what it must mean is a sort of living intelligibility, immanent to life, constituting life's originality and of a unique efficacy. In the final analysis it is destined to be, once it has become a conscious intelligibility within a thinking consciousness, as ours is capable of being, entirely at home at the very hub of the life of the mind, where it is raised to its highest power of truth. The philosopher maintains that this is so true that, by seizing the heart of things in this way, we become at a single stroke indissolubly at one with this meaning.

The man who studies life then sees, to the extent that he goes along with this meaning, springing from the very act of living a kind of arrow of real and progressive energy, becoming more stable as it traverses the billions of years through which life has made its way. This arrow is alive, seeming to accelerate through the various epochs of its system of growth, and at the end turns out to be none other than we ourselves, our conscious collectivity on the march towards its proper destiny of collective conscience. Is this a fair description? You know, at least, that this is what Teilhard de Chardin tried to say, orchestrating it in his magnificent prose. Thus, any meditation on the positive findings of biology must invite the mind to associate with the grandiose universal sphere of relativistic physics the flaming arrow of the meaning of life, so as to create a whole cosmology. This arrow springs into existence with the terrestrial becoming, with the evolution of species and with the appearance of man, and finally acquires its historical task which today still presses onward in its advance. The sphere and the arrow — these are very suggestive symbols! But I must ask again, is this a fair description?

The man who submits to the discipline of physics, particularly classical physics, will, I think, hesitate to say Yes. Seen through eyes accustomed both to the immensity of distant heavens and the immensely small elementary particle, life looks like an episode of a highly irregular sort, taking place in the trifling locality called Earth, perhaps repeatable here and there in the midst of all this immensity, but so far having no known analogue. How can one make a cosmological statement out of one such erratic occurrence? What is more, when he takes a closer look at the facts, our classical physicist grows doubly cautious. He sees life as the fact of a certain category of corporal complexes, once again of a fairly exceptional construction, but made up of elements identical with those found in all the material formations of the universe. We are presently engaged in unravelling the complexities of these constructions and learning the secrets of how they function, and this seems likely to bring us back again to the same architecture and the same physical machinery found in all material things.

When the mind struggles thus with facts of this kind, what happens to the living meaning I was talking about just now, this meaning which is so energetic and real as to be aware of itself? It seems to be of a distressing immateriality, relative to its objective reality and its cosmological contribution. At the very least, the problem is to know whether what we have here is more than a mere subjective delusion, barely rooted in reality, for interpreting the universe, a myth whispered to the mind by Earth, the mother of fables, useful perhaps to mankind but lacking any serious cosmological significance. Should not this whole superstition be banished once and for all so we can get back to the game, the great game of physical, mathematical matter, whose rules apply equally to particles and galaxies, but which does not appear to bestow upon life any particular sort of specificity.

It seems to me, however, that such a way of thinking, though still frequently found, is not quite adequate, even for our physicist. Contemporary physics carries its own future within it, not, it is true, the positive elements of a cosmological theory of life — which must doubtless be left to another range of scientific disciplines — but a whole system of intellectual attitudes which, when all is said and done, seem to suggest that there is nothing absurd, even from the viewpoint of physics, in supposing that the faculties of life and consciousness are indeed one of the essential and constitutive features of our cosmos and ought not to be dismissed as merely an exceptional and adventitious episode. These intellectual attitudes are precisely those of the proofs and concepts that force us to admit that the grandiose representations of the Universe born of relativistic cosmology are still incomplete and unsatisfactory.

It is common knowledge that the theories of general relativity stop short of enabling us to construct an unequivocal model descriptive of the universe. There are a number of representations which look plausible enough on the basis of the data furnished by experience, but this is not enough. Reality confronts us with a variety of general physical phenomena which cannot readily be reconciled with the original contribution of the concepts of relativity to form what the scientific mind calls a "unified field theory", able to assume the universality of any finding. Einstein himself made numerous attempts and many others have since had a go at constituting such a unified field theory, at least for gravitation and electromagnetism, but the results have never been entirely satisfying. Today it is still an intellectual exercise for gifted mathematicians to invent new forms of it. But the mind already seems half-aware that this is not the right way to wave the magic wand of science. Nature is indeed highly sophisticated, but we cannot except her always to speak exclusively in the language of mathematical analysis we have invented.

Indeed, something has happened right there among the physical phenomena to give us a clue to ways of describing and understanding them, ways which are quite different from the classical ones and which could well be utilized for the theory of relativity and unifying projects. The module of physical action has been quantitatively defined. Today the theory of things accessible to our experience obliges us in a relentless and subtle fashion to associate the mathematics of probability with the classic procedures of analysis, although it has not yet enabled us to apply the conventional epistemological justifications for having recourse to the mathematics of probability. It has something else in view. In the person of Einstein, however, the classical spirit strenuously resisted the new ways in which physics, with the quantum theory, was beginning to look at its own body of knowledge and its relation to reality. That debate is still not closed, but the hope that physics may quietly return to the way it saw things before the revelation of the quantum of action now seems rather remote.

To conclude, I am going to put forward a hypothesis which you may perhaps think presumptuous: what the life sciences propose to us directly and positively, as a new and original way of looking at things and at being, is already to some extent heralded and prepared within the conceptual structures presently required by the complex of theoretical developments within physics itself. Clearly, these ideas about life, its indwelling energy and its consistent meaning hinted at by biological reality cannot be made explicit, and they remain in suspense at the level of physics. But this decidedly new feature of contemporary physics, compared with the great theoretical perspectives of classical physics, may be destined to play a double role, on the one hand, as a very essential element in the understanding of reality, even without taking life into consideration, and on the other hand, as a no less essential principle for establishing the cosmic roots of the fact that life exists, sensibility, conscience and all, within the realm of matter.

This would make today's biologist, still an outsider and still half prescientific, a singularly revealing witness of the new concepts with which physics is already pregnant. It may be that one day a cosmology even more scientific than the one we have today may make the originality of the ideas of which Teilhard de Chardin gave such a remarkable illustration seem no longer so foreign and incompatible with the monument of human intelligence whose first rough plan we owe to the wonderful genius of Einstein. This, after all, is what we are here to talk about.

SCIENCE AND SYNTHESIS

Sir Julian Huxley We live in an apocalyptic age, destined to bring forth a new revelation about nature, ourselves and our significance in the cosmic process. To start with, we have to accept the universe, as Margaret Fuller said.

All the reality we know is one vast system of evolution, a directional process tending toward the realization of ever new potentialities. We can study it, within limits we can understand and control its workings; but its and our existence remain as a mystery to be accepted, numinous and compelling, *mysterium tremendum et fascinans.*

Meanwhile, mankind is in travail. Our present psychosocial organization is turning into disorganization and is disintegrating. The urgent need is for a new synthesis — of ideas and beliefs, of social and personal values, of political and cultural aims — to act as a supporting framework and directive matrix for the radically new psychosocial system struggling to be born.

Science is needed as a midwife for the portentous birth. But our present science is not up to the job. Before science can qualify as an evolutionary midwife, it needs discipline and a quick course of training to adjust itself to the new situation. Present-day science needs to synthesize itself before it can be of real service in the necessary task of synthesizing world affairs.

First, however, I want to stress the urgency of the situation. After three centuries of growth and often spectacular success, the present phase of our psychosocial evolution has reached the limits of this usefulness. The system of thought and action in which we all live and move and have our being has become essentially self-defeating and restrictive, and has set mankind on a false course. Today is a time of crisis. If we do nothing about it, 1999 — a date when the majority of the world's present human inhabitants will still be alive — will be a time of disaster.

The four horsemen of our evolutionary apocalypse are, first the military insanity of reliance on the H-bomb as the ultimate deterrent which will prevent nuclear war — though it might better be called the ultimate detergent, as it could wipe the whole of civilization off the face of the earth.

Secondly, the economic insanity of reliance on competitive production and the profit-motive, which is leading to reckless over-exploitation of natural resources, of the customer, and of underprivileged peoples. It is based on a pathetic belief in the possibility of unlimited economic "growth", which implies preferring quantity of material objects to quality of human satisfactions.

Thirdly, the political insanity of reliance on competitive national and bloc power-systems to deal with world problems, which is leading not only to the impotence of what many people are beginning to call the D. N. or Disunited Nations, but also to the widening of the shocking gap between the developed and the underdeveloped or emergent countries — in plain words between the rich and the poor: a gap not only in economic standards but in education and communication, in cultural and recreational facilities, and indeed in the whole quality of existence.

And finally the biological insanity of unlimited human reproduction, which blocks the way to all true progress, and will anyhow make the world of 1999 a much more unpleasant place for our children than that of today is to us. It will wipe out an increasing number of our fellow-creatures, and if unchecked, will lead to intolerable frustration and degradation of life, and eventually to a breakdown of all civilized organization and a return to the armory of pre-human survival — famine, pestilence, and struggle for mere existence.

If we want to label the stage of human development through whose diseased old age we are now living, we must call it the scientific phase.

It began three centuries ago with the realization that organized science, in the sense of unprejudiced, objective and co-operative enquiry, was the key to the acquisition and useful application of new and true knowledge. And it owed its speedy dominance over other organized systems of thought and action, like those based on revealed religions or established hierarchies of class power, to its extraordinary success in unravelling many of the secrets of nature, and then effectively putting them to work in the practical service of man — or at least of some men.

It generated technology (with which it is often confused in the public mind); and in what follows I shall often use the term *science* to include not only pure science but its giant technological offspring. Through technology, science has transferred most of the burden of heavy physical labor from suffering men to insensitive machines; and these mechanical slaves have enormously multiplied the energy that man can command.

This provided the basis for proto-industrialization, and for a huge increase in the volume and the cheapness of goods produced, though at the expense of the army of wage-slaves needed to tend the machines. This led on to mass-production, mass-communication and mass-consumption, with a further staggering increase in volume and cheapness of production and servicing, though too often at the expense of the quality and variety of the resultant products and services.

Today, this continuing trend towards lightening labor and cheapening goods is reaching its inherent limit. It is being pushed to its logical but absurd conclusion. Technology is now providing the basis for automation, which will lead to the abolition of most kinds of work in the customary sense, and so to the threat of excessive leisure (a nice euphemism for under-employment) looming over an increasing number of workers of all classes.

Science so far has given us a vast mass of new knowledge but no knowledge of how to use it. It has indeed become like the sorcerer's apprentice. It has conjured up this dangerous genie we call technology, which is now threatening man's basic ideas about life and how it should be lived. Science must certainly be involved in the most thoroughgoing way in the impending re-synthesis of our collapsing systems of thought and belief, and their organization into a new and better pattern, but meanwhile it, too, is far from being well organized and must do something about organizing itself.

In the first place, there is imbalance between different sciences (or branches of science as they might more modestly be called). This is an imbalance both of prestige and power. Physics, for instance, has the highest status among the sciences. It is regarded by the public — and also, I venture to add, by some physicists — as being in some way more important, more essentially scientific, than biology. It is more exact, we are told, more amenable to mathematical treatment, more rigorously experimental; it has led to a deeper analysis of nature; above all it is more basic.

All this is true; but it is true for reasons which have nothing to do with its relative importance. It is true because physics is the most abstract and therefore the simplest science. It consequently obtained a headstart over the other sciences, grew and developed earlier and faster, and was able to probe deeper and more rigorously into its own area of study just because this is a secluded domain at the furthest remove from the actual complexities of nature, and in particular from the highest complexity we know, the mind-body organism of man. It is credited with probing the mysteries of matter — which are only mysterious until they have been scientifically probed — but it has failed to pay any attention to the only real mystery — the fact of mind and its evolution. In the words of Professor Waddington, the distinguished and many-sided biologist, who, I may remind you, has just completed his term of office as President of I.U.B.S., the International Union of Biological Science, and is now playing an important part in the International Biological Programme, I.B.P., "So long as mind remains a phenomenon unmentioned in the vocabulary of the physical sciences, one can have no complete confidence that the stock of concepts at present used in those disciplines will ever be able to encompass it." After this cautious understatement, he rightly emphasizes the positive duty of science to tackle "all aspects of the phenomena of life, including mind." [1]

It is also true that physics is the basic science and that its laws are fundamental. But we don't regard the basement of our house as more important than the living rooms; and foundations, however essential, are built in order to support a commodious edifice.

The fact is that, at present, there is no such entity as Science, but a number of sciences: different sciences deal with phenomena on different levels of organization and therefore fall on a scale of diminishing abstraction and simplicity and increasing concreteness and complexity (or, as it is more precisely termed, complexification). This has had important consequences. For one thing, we are coming to realize more fully that higher levels of organization can never be fully understood in terms of explanations and principles valid for lower levels. Each new level of organization is emergent in Lloyd Morgan's

[1] Conrad Hal Waddington, *The Strategy of the Genes*, 1957, Allen and Unwin, London, p. 1.

sense, and demands new methods and new concepts. As both Mayr and Simpson have pointedly said, even the most complete knowledge of molecular biology would never enable us to deduce the facts of organismal or evolutionary biology.

There is further the historical consequence that the different sciences have developed in sequence, the simpler and more abstract earlier and faster than the more complex and less abstract. Physics matured before chemistry, and chemistry before physiology; geology matured before biology, and biology before psychology. Indeed psychology has hardly begun maturing: it is still a confused assemblage of warring heresies, and far from being a unified science. And the complexity of anthropological, sociological, cultural, and other human phenomena is such that we have shrunk from imagining a single discipline covering them all: a real psychosocial science is still little more than a dream.

Another consequence is that scientific development has been largely centrifugal; each science has tended to pursue a separate course, exploring its particular demarcated field of study ever more fully. Of course there has been inter-communication, cross-contact, and even cross-fertilization, a kind of reproductive union, producing new generations of scientific offspring, like biophysics or cytogenetics. But one sometimes feels that the separate sciences are behaving rather like galaxies (according to the expanding universe theory), in diverging at increasing rates from some central position towards some limiting frontier.

In any case, there is increasing over-specialization, and uncoordinated scientific information is growing *praeter necessitatem*. What sort of technically improved Occam's Razor is needed to trim this excessive and distorted growth? One useful measure would be to reform the systems of graduate education in general, and in particular to reduce Ph. D. triviality. At the moment, too many Ph. D. theses are burdens on science rather than contributions to its proper development. This reform should go hand in hand with something much more important — the redeployment of science on a centripetal, convergent pattern, instead of its present non-pattern of centrifugal and often divergent trends. This means a concentrated attack on specified problems which involve a number of different branches of science. *Multidisciplinary* is the fashionable word: perhaps plain *cooperative* would be better.

Cancer research exemplifies the value of this way of utilizing our scientific resources. It involves pathology, general biology, embryology, genetics, cytology, biochemistry, biophysics, pharmacology, and clinical medicine, and is now on the verge of a synthesis which will enable us to understand the nature of cancer and in due course to alleviate, cure, or even prevent it.

The astonishing knowledge-explosion of the hundred years since Darwin's *Origin of Species* has produced a greater volume of facts, ideas, and principles than all the previous millennia of history together. This monstrous mass of new knowledge is by no means only the product of the natural sciences: archaeology, pro-history and history, sociology and economics, comparative religion and art history, linguistics, semantics and philosophy, have all contributed large chunks to the pile. Science, or rather scientific cooperation, must clearly play a leading role in the portentous job of synthesizing this vast mass of knowledge into some sort of intelligible, meaningful and humanly effective pattern.

Meanwhile it is true that the knowledge explosion has made possible one important piece of synthesis; it has given us a reasonably comprehensive, reasonably accurate, and readily expansible picture of evolution. (I should strictly say, of galactic evolution, for our picture of cosmic evolution remains obscure so long as cosmologists dispute over their rival theories.)

In any case, it has become clear that all reality *is*, in a perfectly proper sense, evolution — or at least that evolution is a universal process, operating alike in galaxies and stars, in physics and chemistry, in geology and biology, in animal and human behavior.

Briefly, the new picture of evolution shows it operating in three consecutive stages, or phases, the inorganic, the biological and the psychosocial. Each has its own mechanism of operation, its own speed of transformation, and its own type of resultant products. Our own planet is the only place where we know that all three stages have occurred; but on grounds of probability, we can hazard a reasonable guess that this holds for a very small minority of planets elsewhere, and can be pretty sure that the first two stages have occurred on a larger number.

Evolution on this planet can perhaps be envisaged as a tree. Cosmic processes provide the soil out of which it grows; terrestrial physics and chemistry are its roots; biological evolution is depicted by its trunk with its leafy branches; and psychosocial evolution is represented by its flowers and fruits. The comparison is metaphorical and can be misleading. But it may help us to grasp the fact that evolution is the most central and the most comprehensive of subjects and sciences, and that evolutionary science is of special relevance to man's future.

This fact demands practical recognition. If the human species is to avoid disaster, and achieve even a modicum of future success, large amounts of scientific resources, of money, manpower and prestige, must be reallocated from physics and technology to the study of the psychosocial process and the technics of conservation and human improvement. Most of the money and effort expended on outer space research could be much more profitably used on the exploration of the illimitable inner space of the human mind. Most of that spent on hardware and manware for so-called defence, of nations and national ways of life, would be much better devoted to improving the quality of those ways of life and to helping poorer nations. Most of that spent on devising new methods of exploiting natural resources would be better devoted to their conservation, and most of that devoted to encouraging consumer appetite for fashionable and profitably saleable goods, if possible with built-in obsolescence, should be devoted to determining and satisfying man's real needs.

Evolutionary fact also demands recognition. The mechanism of inorganic evolution is physical and occasionaly chemical interaction; and it operates with extreme slowness. Its small-scale products include atoms and simple chemical compounds, its large-scale products galaxies and stars.

The mechanism of biological evolution is natural selection; it operates more rapidly; and its products are organisms, species, large-scale groups, and ecological communities.

In human evolution, the chief mechanism is psychosocial selection working through cultural pressures; it operates much more rapidly and with accelerating speed; and its results are human beings living in societies, with their concepts and beliefs, their economic and political organizations, their laws and their religions.

The three phases are separated by two critical thresholds, the first involving the emergence of self-reproducing and self-varying matter, the second that of self-transmitting, self-varying and self-correcting mind, with cumulative cultural traditions as a result. Each leads to a higher type of organization, and does so by combining a series of steps in the right direction.

The same sort of thing is needed within each phase, to surmount similar crises and to achieve the passage from lower to higher types of organization and achievement. These too are rare, and must be achieved stepwise by synthesizing a number of improvements. Such was the critical passage of vertebrate life from water to air, or that of human life from food-gathering and hunting to civilization: and such is the present crisis involving the passage from an outworn system to a future which has not as yet taken shape.

But I must return to the fact of crisis. Any scientist who emerges from the safe rabbit-burrow of his special subject and takes a good look at the world's general situation is virtually forced to assume the mantle of a prophet. He prophesies doom; but, being a scientist, the doom he prophesies is a conditional doom: he prophesies doom — *unless* we do certain things; if we do them, doom is avoided; and he can become a prophet of destiny.

I must first specify doom in a little more detail. Our nuclear insanity is bound up with the inadequacy of our brains. Our brains were evolved to react to and cope with relatively small-scale and immediate problems: they are simply unable to react meaningfully to problems of a much higher order of magnitude. As Szent-Györgyi has recently said, "with my brain, adapted to handle a primitive fire in a cave, I cannot imagine a fire of 15 million degrees or an atomic blast 300 miles away which would blind me for ever." Having been evolved to live with a small clan, man is still touched by individual and local suffering; but he cannot multiply individual suffering by a hundred million, and so we talk glibly about the pulverization of cities and the "overkill" of a hundred million human beings. Quick and progressive reduction of nuclear weapons is essential, and has immediate priority as a doom-avoiding measure.

But population increase will remain as a graver threat. Even if the nuclear threat is dealt with, over-population remains as the gravest long-term threat to human progress, and the one most difficult to cope with, as it is rooted in man's physiological nature.

In sum, unless we abolish the threat of all-out nuclear war, we risk the doom of destroying human civilization and of saddling the human species with grave genetic damage and multiple deformities. And even if we avoid nuclear war, then, unless we stop the increase of human population, we confront the risk of becoming the cancer of our planet and the certain doom of debasing civilization and destroying the resources needed to restore it. In the one case we give human evolution a violent set-back, from which it might never recover; in the other we reserve its direction, from advance to retreat, from progress to regression.

A gloomy prospect! But I believe that we can take the measures needed to nullify these major threats. After all, the very existence of man is bound up with his self-regulating and self-correcting cybernetic capacities. He can become aware of his errors and then correct them in the light of his past experience and with the aid of his accumulated knowledge.

Even so, the general crisis remains — the inadequacy of current religions and ideologies, the confusion of general thought, the wretched multitudes of the poor countries making shift to live with their miseries, the discontented multitude of the affluent countries with their massive indifference and their sporadic outbursts of stupid violence, the absence of any guiding beacon in this dreary twilight of existence, the lack of any overriding belief to give meaning to men's lives and inspire them to action.

Scientifically speaking, what we need is a better model of man and his evolutionary capacities — in more human terms, a new, integrated and compelling vision of man's destiny. Only with its aid shall be we able to surmount the present crisis and enter a phase of human renewal, and for its patterning we need both science, in the broad sense of objective, organized and usefully applicable knowledge; and also synthesis, in the broad sense of an attempt at integrating scattered facts and conflicting ideas into a meaningful whole.

The outlines of this vision are already emerging. Man's destiny is to be the world's senior partner, the primary agent for the future evolution of this planet. This applies both individually and collectively, both in the short and in the long term. In him, the evolutionary process has finally become conscious of itself. This is a unique privilege, but also a formidable responsibility, which gives him dignity, but which he cannot unload on to the shoulders of God or Fate.

The new vision also reveals man as young and very imperfect, but with a vast stretch of time in which he can realize his immense stock of potentialities. It provides both encouragement and hope. First of all, evolution on this planet is directional. During the three billion years of its biological phase, it has moved steadily towards the improvement of life, towards better balance, greater variety and higher organization — first of body, then of mind. Man is the latest product of a continuous process of biological improvement, punctuated by critical steps from one dominant type of organization to a new and higher one — from a sub-microscopic speck of self-reproducing matter, through cell and complex invertebrate, increasingly sentient amphibian, reptile and mammal; and on to the incredible complexity of conscious man, with more cells in his brain than there are people in this world.

A similar process has been at work in man's psychosocial history. But here the improvements are cultural, and the dominant types are not organisms but psychosocial organizations of ideas and beliefs with their attendant social structures and economic practices.

Among the obvious critical points in psychosocial evolution are the passage from food-gathering to organized hunting; that from hunting to agriculture and settled life, and on to early civilization; and the later crisis of the Renaissance and the Reformation which led to the emergence and eventual dominance of a science-based system.

According to the second law of thermodynamics, the entropy of the physical world is increasing, its level of organization is getting lower. The world of life, however, with the aid of solar energy, is pursuing an anti-entropic course. Is level of organization has steadily risen, but increasingly fast with the evolution of mind; for mind makes it possible to achieve astonishing organizations, both of thought and its applications, with a lower expenditure of physical energy. It should be noted that in information theory, organization is equated with information; to which I would add that higher organization possesses greater significance in the context of ongoing evolution.

Yet the scientific study of mind has barely begun. Who can doubt that its full exploration will lead to unimagined heights of mental and spiritual organization, with undreamt — of possibilities of rewarding experience and relevant action.

Coming down to the immediate future, it is clear that any new synthesis must be open-sided, and that its aims must be stated in terms of fulfillment — greater fulfillment for more human individuals, fuller achievement by more human societies, a greater sense of personal significance, more profound involvement in the evolutionary task of cooperative world development in all its aspects, and a radical shift in values from quantity of objects and products to quality of life and living. We must replace the idea of the Welfare State by the Fulfillment Society. In preparation for this, our educational systems should be organized so as to provide individual fulfillment through adventure and exploration, through physical and intellectual participation in worth-while projects of study and service, and the enjoyment of love and beauty.

We should also start planning for social fulfillment through sane population policies; through creating cities which would be organs of civilized life instead of generators of frustration; through balanced land-use giving opportunities for solitude and enjoyment as well as for industry, building and communication.

We must start thinking about how production could be properly organized so as to leave people free to devote most of their time of learning, teaching and aiding, to the enlargement of experience and the achievement of a greater sense of significance.

Most people are prisoners of their conflicts and their own little selves; their primitive ways of thinking saddle them with guilt and make them project own repressed nastiness on to others. Already we know of various methods of achieving wholeness and self-transcendence: we must push ahead with the development of psychotechnology on a large scale. My brother Aldous in his last book, *Island,* has outlined various of the methods by which human beings could find fulfillment and wholeness and enjoy richer experience.

On the world scale we must give effective expression to the central fact of the unity of man, while encouraging cultural variety. We must have a world population policy, and try to link all nations in unified projects of world development.

I must end with one particular plea — that Unesco continues to promote the great debate which the Director General has today initiated. The building of a new organized system of thought and belief is the greatest task of this generation. I hope that Unesco will organize continuing discussion of the subject in all its Member States, in the world's main regions, and of course here at Unesco House. In so doing, it would emphasize its rightful function of intellectual, cultural and moral leadership in the realm of the human mind.

THE GREAT COSMOLOGICAL DOCTRINES

Giorgio de Santillana I am going to walk straight into my subject by the front door, the monumental portal of ancient Greece. We see before us the severe and spare architecture of the pronouncements of Anaximander, some 600 years before Christ — spare it most certainly is, for we possess only three or four phrases of his. "That of which all things are born is also the cause of their end, *as is just,* for they must make amends and expiate the injustices they did to one another in the order of time."

All things are born and die in a universal rhythm, and so the first term of the equation is a law of periodicity. The second term — expressed in "*as is just*" — is a reminder of indwelling justice and an invocation of the sovereign order of time. This is the very beginning of physics, with its opposing forces, its dynamic equilibrium which turns into a whirlpool and sets off the evolution of life, as is explicitly stated. Aristotle later forgort this idea and it did not re-emerge until Leonardo da Vinci.

The substance of the whirlpool, a model of the spiral nebulae, is *Apeiron*, the boundless. This substance is also said to be the Divine, the *Arché* of all things — a word of pregnant meaning, if ever there was, signifying primacy, supreme power, principle and beginning.

He goes on: "The Boundless understands and rules all things." The word used here is *kybernai*, to direct or steer. This rule is conceived, not as a voluntary action, but as self-regulating, which should be the case in a well-conceived society. This is an idea truly physical in its conception, although Animaxander did not have the words to describe it. These words are: automation and cybernetics — as you can see, it was all there in germ!

And here, at a single stroke, a real physical principle emerges. Why does the Earth not fall down? Because it is located symmetrically relative to its surroundings. There is no *sufficient reason* why it should fall down.

This is clearly the first draft of the principle of *sufficient reason* which Mr. Oppenheimer quoted when he was talking about Einstein. What a strange and wonderful beginning to come out of a purely vitalistic world! Under this particular negative aspect, this is the principle which releases the mind from good reasons in our own image and on our own puny scale and launches it into the immense adventure of science unbounded. A once-only chance, a throw of the dice which stopped men from dreaming about organisms and moved them to draw up their first model in terms borrowed from their technology: winds and coulds, sieves and whirlwinds. It was this model which steered them towards new horizons and new and burning questions — that intellectual acuity which is called *Sophia*.

Here is the point of origin of *scientific rationalism*, an intellectual fever which can declare that what one thinks without contradiction must be able to be true, however strange and new it may be.

On this foundation there arose another dazzling novelty, Pythagorean mathematics, hand in hand with the creation of the ideal of culture which the Greeks called *mousikè*, which implies a combination of logos, melody and movement. So here we are already in the open sea of naked, materialistic physics, with the true powers of nature defined as harmony and proportion.

Number is on the march, an autonomous force which is not an element like air or water, but a genuine *principle* with a life of its own; and this is right and proper for it represents the laws of thought.

In this way physics itself becomes *mousikè* and astronomy the royal art it always had been. Pure form and periodicity, these are the main themes laid down by geometry, and they led Aristarchus, 270 years before Christ, to propose the Copernican system. And so our cosmos was born, of which Kepler

2000 years later sings in the language of a prophet: "Behold, a light appeared unto me, the sun of truth. Nothing can now hold me back; the die is cast. I write for posterity. The Lord waited 6000 years for his works to be understood; I too can wait."

This is his hymn to the harmony of the stars. But soon after this the harmony of heaven and earth was to be made discord; mechanics was to be born and Descartes would destroy the cosmos. Then, when the reign of Newton set in, the universe became a carefully wound clockwork mechanism and God a despot playing with his mechanical toy.

Later still, things were to become more complicated; Mr. Holton is going to tell us about that, but this is where the idea of space has brought us, to the point where it receives philosophical affirmation in Einstein's total geometrization.

Now let us go back to the beginning. Of Anaximander's four great phrases, two look forward, predicting modern space-based cosmology, and two look back to the traditional idea of the sovereignty of the Order of Time. For the fourth, which comes down to us through Cicero, states that, according to Anaximander, "the gods have their beginning at long intervals in the East and in the West, and they are numberless worlds." These strange words, of which Cicero said they were Greek to him, can be understood only in relation to an almost forgotten cosmology which comes down to us from the dawn of time and which survives only in myths and fables. It is the vestige of a technical language from before the invention of writing and which we are painstakingly trying to reconstruct. I shall therefore go back even further to look for the remnants of these beginnings.

Would you like me to show you the thing in its primitive state? Here is a story from the Red Indians of the Canadian Pacific coast, one among thousands, picked almost at random, and to which nobody has paid much attention. It derives from the Satlo'lq Indians of the south-east of Vancouver Island.

"Once upon a time there was a man whose daughter had a magic bow and arrow, but she was lazy and did nothing but sleep. Her father grew angry and said, 'Instead of sleeping, take your bow and try to hit the navel of the ocean, so we can have fire.' Now, the navel of the ocean was a big whirling funnel and there were bits of wood floating in it for making a fire. For at that time there was no fire. The daughter took her bow, drew it and hit the target and the sticks flew on to the shore. The old man was pleased and lit a fine fire. But he wanted to keep it for himself, so he built a house with a door which fell shut from top to bottom, like a jaw, and it killed everyone who tried to go through it.

But the news went round that the old man had fire, and the Stag decided to steal the fire for the people. He stuck bits of resinous wood in his hair, then he bound two canoes together and built a bridge of planks between them, and he sang and danced upon it as he paddled towards the old man's house. He sang,

'Ho, I am going to catch fire.' The old man's daughter said, 'Let him in, I like his song,' but the door of the house fell shut. Guided by his song, the Stag reached the door just as it was opening again and got inside without harm. There he leaned over the fire as if to dry himself, and the twigs in his hair caught fire. Thereupon he leaped out of the door, and this is how he brought fire to men."

This is the Prometheus story in Satlo'lq. But it is much more as well. The Stag is not merely our heroic Prometheus (as described by Aeschylus and Shelley, just to get the record straight); he is Kronos, Saturn, the greatest of the earthly gods, the Cosmic Demiurge, Deus Faber. In the Hindu tradition, Kronos is called Yama and has a stag's head. This animal's head is found right through the ancient world. Now, if you care to look — only no-one ever does — in the Orphic hymns, you will find that Number 13 in the old Hermann edition is a hymn to Kronos: "O thou, almighty devourer, ever-reborn Kronos, great Aiôn, venerable Prometheus" — in Greek, *Semnè Prometheu.* Now, I did not put these words into their mouths.

And Sophocles' Scholiast, quoting the lost sages, Polemon and Lysimachides, says that in the gardens of the Academy there was an altar dedicated to Prometheus, "first and most ancient holder of the sceptre", and to Hephaïstos, "the second and younger". Now, those of us who are in the business know that Hephaïstos appears as the second aspect of Saturn, or more specifically Demiurge, Deus Faber. But it is to the grim Kronos, the all-knowing, "the planner", that the Promethean prescience goes back. We are here in the basement of Greek antiquity and, by a strange subterranean route, it needed a primitive people from the other side of the world to draw our attention back to certain Greek texts which throw new light on the classical myth.

As for Omphalos, the navel, there are whole books of studies on this. It is the island of Calypso, but also Charybdis of the Odyssey, the eye of the whirlpool (Maelstrom) of Indo-European tradition, the *gurges mirabilis* transpiercing the world and coming out in the Land of the Blessed which, of course, is in the southern heavens at Eridu, or else the ship Argo where sleeping Kronos is king; for the Hindus, Yama Agastaya; for the Egyptians, Osiris, judge of the dead; for the Babylonians, Ea-Enki; for the Mexicans, Quetzalcoatl — and many more. This, if we are to believe Dante, is where Ulysses went astray; it is where Gilgamesh lies seeking immortality at "the confluence of the celestial rivers".

And what about those bits of wood in the whirlpool? This is the *other* whirlpool, the cosmic one, the Precession of the Equinoxes which was known even then and which, over a span of 26,000 years, establishes the Order of Time. This is where the primitive Prometheus is at home (Pramantha in the Indes), and the fires are not those of midsummer but of the passing of the equinoctal sun from one sign of the Zodiac into the next, which happens roughly every 2,400 years; the end of one "world", or era, and the beginning of the next.

I would like to quote a line from the French poet, Agrippa d'Aubigné; he is writing about the end of one of these worlds:

> "... quand les esprists bienheureux
> Dans la Voie de Laict auront fait nouveaux feux ..."

(when the Spirits of the Blessed will have kindled new fires in the Milky Way.) This is the moment when, in Mexico, Tezcatlipoca lights a new fire by twirling his stick in the sign of the Gemini and "from this time on he was called Mizcoatl".

After this, things become more complicated (I must ask you to be patient a while longer); the original fire of Mizcoatl was supposed to have been lit at the Pole, so it is not at all clear why it happened in Gemini at the same time, but this is attested by a number of ceremonies. One should perhaps see in this a kind of ambivalence or bilocation of the sacred fire which consecrates the equinoctal colure of this famous year zero, from which time was counted in Mesopotamia, in China and in Mexico, too: the solemn moment when the sun of the spring equinox passed under the sign of the Gemini and so, too, on to the Milky Way, and the great galactic arch (or bow) raised on the horizon made an almost complete circle, the equinoctal colure. This, indeed, marks the fundamental geometrical scheme of this cosmogony, as it is found more than once.

Our story even has its proto-Pythagorean element. The rhythm of the Stag's singing and dancing becomes in another story from northwest Canada (British Columbia, Lower Fraser River) the prowess of the grandson of the Woodpecker who, on the point of drawing his bow, intones a song and, once he has found the correct note, the arrows which he shoots thrust one into another to form a bridge between earth and heaven. This is a truly Orphic theme, many times repeated, but also, as Sir James Frazer himself remarked, a lingering memory of the scaling of Mount Olympus in the Gigantomachia.

So now you see where such a story, which appears to have neither head nor tail, can lead us.

And yet, you will say, what a confusion of ideas! I quite agree. Still, we have to remember that their transmission was exposed to all the hazards of being passed on by word of mouth, of being forgotten and of being misunderstood. The very disorder only accentuates the authenticity of the component parts, the incredible pertinacity of certain ideas in surviving — surviving indeed their own meaning — like a sacred trust from lost ages.

To continue, the bow and the arrow are constantly reaffirmed as basic images, keystones of the arch of theory; both are found in the heavens: the bow of Marduk, the Babylonian Jupiter, the bow celebrated by the Poem of the Beginning, with which he won power and established universal order. Then again, it is the bow the Chinese emperors received on their accession. With this bow one cannot fail to "hit" Sirius, "he" — so says the great Babylonian ritual of Akitu — "who measures the depths of the sea". There have been many books written about this, too. But the Schlegels, the Guérins, the Gundels and all the great scholars who by their prodigious labors have elucidated this uranography, tend to be imprisoned each within his own province, be it Mesopotamia, India, Egypt or China, and each has instinctively claimed the privilege of discovery for his protégés, leaving to others the labor of building a bridge between these diverse civilizations. There have also been distinguished astronomers men like Biot and Henseling, whose books are no longer read and whose efforts at comparison have run away into the sand.

It was, however, by repeated cross-comparisons that the meaning of these enigmatic words was ferreted out. The star, Sirius, has been an object of fascination in many latitudes and frequently obscure allusions are found to its links with the sea — right down to Aristotle and Pliny. Sirius seems to have been a kind of pivot for a number of intersecting lines drawn from various regions of the heavens. The main alignment places Sirius on the line joining the Poles, terminating in the south at Canopus. This is another very fascinating star, the seat of Yama Agastaya for the Hindus, the mythical city of Eridu for the Sumerians and for the Arabs Suhayl-the-Heavy, because it marked the bottom of the "celestial sea" of the southern hemisphere. The other alignments linked Sirius with the "four corners of the heavens", equinoxes and solstices which owing to the Precession were moving imperceptibly throughout the centuries; the line of the North Pole passed over all the stars of the Great Bear in turn, like a needle moving round a huge dial.

It seems that these angular measurements were solemnly and carefully checked on high days and holidays. It was conceived that by means of Sirius the earth was indeed "anchored to the depths of the Abyss" and "hitched" to the southern heavens; through this star it was possible to check whether the universe was functioning properly. This, so far as we can divine it, was the mythical and ceremonial role of the Bow of the gods.

Thus, the only originality of the Redskins seems to have been to place the bow in the hands of a woman, and a lazy one, at that! Is this a faint echo of Ishtar, the cajoler? I would prefer to believe that the Indian storyteller had his moment of invention. There are, after all, so few in this rigidly traditional poetry!

To return to the door which shut like a trap, this is quite in keeping with Homer's *Plangtai,* the Simplégades of the Argonauts, the rocks-which-crash-together, or even further back, in the primitive vertical figuration, it could be the Ecliptic, rising and falling on the horizon in yearly cycles, the object of innumerable parallel figurations spaced out over the continents since at least the fifth millenium. But then savages, of course, hadn't a clue about astronomy — you are quite right to remind me of this fact. True enough, they hadn't any more, and the people who did have this knowledge were not savages, no more than the builders of Stonehenge, although until last year the archeologists persisted in calling them "howling barbarians". But last year Gerald Hawkins, a young astronomer who knew his earlier colleagues by their works, came up with his ordinators. Sir Norman Lockyer had quickly been buried and forgotten, because philologists have very little in common with astronomers, but he has now had to be rehabilitated.

I have offered you here a whole encyclopedia full of themes condensed into a dozen lines. I have made you a gift of one of those Japanese flowers, creations of folded paper which open into enormous patterns when you drop them into a bowl of water. The first opening of the Japanese flower can be seen in Lycophron's *Alexandra,* a conjuring poem of the Archaic myth; it opens a little more with Apollodorus' *Bibliotheca,* that reservoir of classical myths, and in Ovid's *Fastes,* the commentaries of Proclus and Porphyrius, and comes to its full blossoming, or almost, in the *Dionysiaka* of Nonnos the Panopolite, or the *Tetrabiblion* of Ptolemy, which are the true manuals of Archaic mythology.

Now everyone knows that the innovator was Sir James Frazer, that tireless seeker after pre-classical antiquity; but, although he had a well-tended garden, with all the myths laid out in formal beds, he was not able to decipher their meaning, for he ignored the vegetation of the cults. There was, to be sure, a little hut tucked away in one corner where he could have found the tools and the plans, but Sir James didn't notice it. He was a full-blooded evolutionist and for him everything had to be either agriculture or magic.

I do not wish to detract from the virtues of Cook or Jane Harrison and all their school, but if they had had fewer pre-conceived ideas they might have found, in addition to the annual cycle recorded in the vegetation, other cycles of 2, 4, 8, 12, 30, 52 and even 60 years which had quite a different significance and could only be planetary cycles. But there it is, in all the excitement of a new idea a lot of things are overlooked and the problem of dealing with the loose ends is left until later.

I have given you an example of how a universal mythical language, predating written records, has been found to conceal a lost cosmology which was also universal — quite as universal as the one we have today. Not the least of the mysteries involved is how these ideas traveled by diffusion from protohistoric Mesopotamia, since it seems to have been here that the planetary cults — which Aristotle attributes to the *panpalaíoi,* the most ancient — and proto-Pythagorism, too, first came into existence.

What was diffused clearly was not ready-made ideas, but patterns: the Ecliptic and its constellations, the positions of the heavenly bodies, the celestial zones, certain key myths and all this curious urano-geography where heaven and earth meet under the domination of the sovereign planets in their inexorable courses. Then again, it was the link between harmony and the heavenly bodies, harmony and units of measurement, the sovereign principles of precision, called *maat* in Egypt, *rta* or rite in India. "Between the music of the ritual pipes and the calendar" — says a Chinese proverb — "the adjustment is so fine one could not draw a hair between them." And in just this way alchemy was adjusted to astrology, then astro-medicine, plants, metals, alphabets, learned games like chess, magic squares like the one recorded in Dürer's *Melancholia* — the microcosm adjusted to the macrocosm. The whole thing was not set out like a system of logic, but like a fugue in music, as befits a genuine self-regulating organism. This world was not merely determinist but pre-determined — on a number of interacting levels; it was a world supersaturated with determination, where total necessity is queen but freedom yet remains, as with Spinoza's God. "And they say," remarks Aristotle the Modern without too much charity, "they" being the Pythagoreans, "that the interval between the letters alpha and omega is the same as that between the top and bottom notes of the *aulos,* and that the number of them is equal to the heavenly choir." Thus, these ideas persisted right into classical times, thanks to the fervency of the Pythagoreans. We still retain from it number and rhythm, the incidence of the unique moment, the idea that the time is ripe, the *kairos,* as the Greeks called it, which decides between being and non-being. There was a time when justice was above all *justness* and to be inexact was a sin.

Since Descartes, we moderns think in terms of simple space, something we can dominate and where our actions are recorded. Archaic man thought in terms of time dominating all else. And indeed, what we persist in regarding as "distances" in his system are really angular measurements which change with time. Spatial order as we understand it is meaningless; for him, it was modulated entities, intervals on a vibrating string, spheres, triangles, magic squares, polyhedra. Even for Plato, pure space, what we would call Newtonian isotropic space, was the nearest thing to non-being. Parmenides himself was unable to give his Being a being other than by assigning him bounds in the form of a sphere. It was said in China, "The sovereign rules over space because he is master of time."

This is the way mankind thought during as many centuries as separate us from the Great Pyramid. This thinking was totalizing, if we may borrow the term from Lévi-Strauss. The Order of Time, which was the true cosmic order, governed the fate of lives and of souls. It supplied an eschatology as well as a science to generations without number in the dim and distant past. If we now seek to recover it, the reason is that we hope to restore their true shape to civilizations long buried and forgotten; whole contintents long classed as having no history are now stepping on to the world stage, armed with fresh claims and ready to reassume their proper role in the history of our species.

Our ancestors said to themselves, like Cocteau, "Since these mysteries are beyond us, let us make ourselves organizers of them." What we have here is thrilling material, only just beginning to be deciphered; however, our generation has brought up reinforcements, names like Hartner, van der Waerden, von Dechend, Needham, Werner, Marius Schneider, forces assembled from all points of the intellectual horizon.

What we need now is a convergence of ideas, disciplines and methods, as well as the aesthetic intuition to enable us to unravel this Art of the Fugue and to comprehend, as d'Alembert once said, "our forefathers to whom we owe everything and about whom we know nothing".

WHERE IS REALITY? THE ANSWERS OF EINSTEIN

Gerald Holton To do justice to any part of Einstein's work is of course not possible in a talk. But on an anniversary occasion such as this, one can do two things. First, one can celebrate a great and good man; and while Einstein has no need of our celebration, and was always rather amused by this sort of affair, we are right to do it because it is good for us to remember him. And secondly, one can try to add something to what is already known, thereby making an offering, a wreath of flowers from one's garden. This is what I want to present here today — a chapter in the history of ideas that illuminates Einstein's pilgrimage from a philosophy of science in which positivism was the center, to one in which the center was a rational realism. To the extent that it is possible without unduly neglecting the major published contributions on this topic, I hope to use here the unpublished scientific correspondence and other documents in the Archives of the Estate of Albert Einstein at Princeton which I have recently been studying and helping to order for scholarly purposes. [1]

1. *Philosophical Background*

If one studies Einstein's earliest papers on special relativity from a philosophically critical point of view, one can discern the influence of many, partly contradictory, points of view — not surprising in a work of such originality. There are debts to Hume and Kant and Poincaré, for example. But perhaps the most important of these influences was that of the empirio-critical positivism of the type of which the Austrian philosopher Ernst Mach was the foremost exponent. Let us remember the philosophical background of that period. The 1890's and 1900's were a time of turmoil not only in physics but also in philosophy of science. There were vociferous opponents of kinetic, mechanical, or materialistic views of natural phenomena. They objected to atomic theory and gained great strength from the victories of thermodynamics from Carnot on, in which knowledge or assumptions about the detailed nature of the material (e. g. in understanding heat engines) were not needed at all.

In the late nineteenth century the critics of the mechanical interpretation of physical phenomena included W. Ostwald, Helm, Stallo, and Mach. Their positivism provided an epistemology for the new, phenomenologically based

[1] I wish to acknowledge with gratitude the help received from the Trustees of the Albert Einstein Estate, and particularly from Miss Helen Dukas. Permission to quote from the writings of Albert Einstein may be sought from the Executor of the Estate.

science of correlated observations, linking energetics and pure sensationism. Ostwald, in the second edition of his great text on chemistry, in 1893, gave up the mechanical treatment for Helm's "energetic" treatment of the subject. Here, as in the work of other energeticists and phenomenologists, "hypothetical" quantities such as atomic entities were omitted; instead, these authors claimed they were satisfied, as Herz wrote around 1904, with "measuring such quantities as are presented directly in observation, such as energy, mass, pressure, volume, temperature, heat, electrical potential etc., without reducing them to imaginary mechanisms or kinetic quantities".[2] As a consequence, the introduction of such conceptions as the ether, with properties not accessible to direct observation, was condemned. Instead, these philosophers urged a reconsideration of the ultimate principles of all contemporary physical reasoning, notably the scope and validity of the Newtonian Laws of motion, the conception of force, and the conceptions of absolute and relative motion.

At the close of the nineteenth century Ostwald wrote, "Energy is a real thing, indeed the only thing in this so-called external world"; he objected to those who held that the assumption of that medium, the ether, is unavoidable. "To me it does not seem to be so ... There is no need to inquire for a carrier of it when we find it anywhere. This enables us to look upon radiant energy as independently existing in space." This spells out prophetically the kind of suppositions behind the Einstein papers of 1905 concerned with the photon theory and with relativity theory.[3]

Ostwald's main philosophical ally was Ernst Mach. His great work, *The Science of Mechanics,* published in 1883, is perhaps most widely known for its discussion of Newton's *Principia* — in particular for its devastating critique of what Mach called the "conceptual monstrosity of absolute space" (Preface, 1912 Edition), a conceptual monstrosity because it is "purely a thought-thing which cannot be pointed to in experience".

Starting from his analysis of Newtonian presuppositions, Mach proceeded in his program of eliminating all metaphysical ideas from science. As Mach

[2] J. T. Herz, *A History of European Thought in the Nineteenth Century,* New York, 1965 (reprint, Dover Publishing Co.), vol. II, p. 184.

[3] Documents have been found recently to show that Einstein esteemed Ostwald. Einstein's first publication (1901), on capillarity phenomena, used data published by Ostwald, and indeed was inspired ("angeregt") by Ostwald's work, according to Einstein's letter of 19 March 1901, to Ostwald (published by F. Herneck, *Forschung und Fortschritte, 36,* 1964, p. 75). The reason for Einstein's letter was that he had failed to receive an assistantship at the Eidgenössische Technische Hochschule in Zurich; he turned to Ostwald to ask for a position at his laboratory, partly in the hope of receiving "the opportunity for further education". Not having received an answer, Einstein wrote again to Ostwald on 3 April 1901; and on 13 April 1901 his father, Hermann Einstein, wrote Ostwald a moving appeal, evidently without his son's knowledge. Hermann Einstein reported that his son holds Ostwald "most highly among all currently active scholars in physics". The only other known attempt on Einstein's part to obtain an assitantship at that time was a request to Kamerlingh Onnes (12 April 1901), to which he also seems to have received no response.

said quite bluntly in the preface to his first edition, "This work is not a text drill in theorems of mechanics. Rather, its intention is an enlightening one — or to put it still more plainly, an anti-metaphysical one."

Mach's influence was enormous. Although he repeatedly insisted that he was beleaguered and neglected, and that he did not have, or wish to have, a philosophical system, his philosophical ideas and attitudes had become so widely a part of the intellectual equipment of the period from the 1880's on that Einstein was quite right in saying later that even Mach's opponents were unaware how much of Mach's ideas they had, as it were, "imbibed with their mother's milk". The problems of physics themselves helped to reinforce the appeal of the new philosophical attitude. The main problem was to reconcile the notions of ether, of matter, and of electricity be means of physical pictures and hypotheses. An example was Larmor's definition of the electron as a permanent but movable state of twist or strain in the ether which formed the atoms of electricity and possibly ponderable matter itself. Larmor so tried to bridge the continuity and uniformity of ether and the discontinuity of particles of matter and electricity.

The unease and impatience with this kind of physics was getting very marked in some quarters at the turn of the century. No matter how some of the newer physicists of the time wrestled with the problems of physics, the use of conceptions developed in nineteenth century physics seemed to them merely to produce failure and despair. It is not too much to say that the new physics they fashioned arose, first of all, from this experience of despair. And here the role of Mach as iconoclast and critic of classical conceptions was particularly important, for whether or not they correctly assessed or valued Mach, the critical force and courage exemplified in Mach's thinking made a strong impression on them.

2. *Mach's Early Influence on Einstein*

As the Einstein correspondence in the Archives at Princeton clearly shows, one of the young students of science most caught up by Mach was Einstein's closest friend and fellow student, the only one whom Einstein credits in his 1905 paper on relativity with having been helpful to him, namely Michele Besso. You will remember that Einstein noted in his autobiographical essay (1949) that Ernst Mach's *History of Mechanics* "shook this dogmatic faith" in "mechanics as the final basis of all physical thinking ... This book exercised a profound influence upon me in this regard while I was a student. Mach's epistemological position influenced me very greatly." In a letter of 8 April 1952 to Carl Seelig, Einstein wrote, "My attention was drawn to Ernst Mach's *History of Mechanics* by my friend Besso while a student, around the year 1897. The book exerted a deep and persisting impression upon me ..., owing to its physical orientation toward fundamental concepts and fundamental laws. Only Michele Besso and Marcel Grossmann were close to me during my period as a student."

It was Besso, writing in 1947 to Einstein, who still said, "As far as the history of science is concerned, it appears to me that Mach stands at the center of the development of the last 50 or 70 years." It was Besso who wrote to Einstein, in December 1947, to remind him of the consequence: "Is it not true," Besso asks, "that this introduction [to Mach] fell into a phase of development of the young physicist [Einstein] when the Machist style of thinking pointed decisively at observables — perhaps even, indirectly, to clocks and meter sticks?" [4]

We now can ask whether and in what sense Einstein's relativity paper of 1905 was, to a large extent, Machist — not counting the characteristics of "clarity" and "independence," the two traits in Mach which Einstein always praised most. The Machist component of the 1905 paper shows up prominently in two respects. First, by Einstein's insistence from the beginning that the fundamental problems of physics cannot be understood until an epistemological analysis is carried out, most particularly so with respect to the meaning of the conceptions of space and time. And secondly, Einstein's first relativity paper clearly has a Machist component because he equates reality with the givens, the "events," and does not, as he later did, place reality on a plane beyond or behind sense experience.

Mach's program was later characterized by Einstein himself in a brief and telling analysis, published in the *Neue Freie Presse* of Vienna on 12 June 1926, the day of unveiling of a monument to Mach. Einstein — then already disenchanted from some years with the Machist program — wrote: "Ernst Mach's strongest driving force was a philosophical one: the dignity of all scientific concepts and statements rests solely in isolated experiences *(Einzelerlebnisse)* to which the concepts refer. This fundamental proposition exerted mastery over him in all his research, and gave him the strength to examine the traditional fundamental concepts of physics (time, space, inertia) with an independence which at that time was unheard of ... Philosophers and scientists have often criticized Mach, and correctly so, because he erased the logical independence of the concepts vis-à-vis the 'sensations,' and because he wanted to dissolve the reality of Being, without whose postulation no physics is possible,

[4] Many other evidence have been published to show Mach's well-known influence on Einstein during approximately the first 15 years of this century. An interesting point that seems to have not been generally noticed is the report that recently a document has been found which shows that Mach in 1911 had participated in formulating and signing a manifesto calling for the founding of a society for the positivistic philosophy. Among the signers, together with Mach, we find J. Petzoldt, David Hilbert, Felix Klein, George Helm, Sigmund Freud. And also Einstein, who at the very time was in correspondence with Mach himself. (See F. Herneck, *Physikalische Blätter, 17*, 1961, p. 276.)

in the reality of experience." But this was written in 1926, and anticipates our story [5].

Now we return to Einstein's 1905 paper on relativity theory *(Ann. der Physik, 17,* 1905, pp. 891–921). From the beginning the instrumentalist view of measurement and of the concepts of space and time is made to appear most prominent. The key concept in the early part of the paper is introduced at the top of the third page in the following, straightforward-sounding way (indeed, it has been said by Leopold Infeld to be "the simplest sentence[s] I have ever encountered in a scientific paper"). Einstein wrote: "We have to take into account that all our judgements in which time plays a part are always judgments of *simultaneous events.* If for instance I say, 'that train arrived here at seven o'clock,' I mean something like this: 'The pointing of the small hand of my watch to seven and the arrival of the train are simultaneous events.'"

The key concept introduced here is the concept of *events* — a word that recurs in Einstein's paper about a dozen times immediately following this citation. Transposed into Minkowski's useful scheme, Einstein's "event" is the intersection of two particular world lines, say that of the train and that of the clock. The time (t co-ordinate) of an event by itself has no operational meaning. As Einstein says, "The 'time' of an event is that which is given simultaneously with the event by a stationary clock located at the place of the event ..." And just as the *time* of an event assumes meaning only when it connects with our consciousness through sense experience — that is, when it is subjected to measurement-in-principle by means of a clock present at the same place — so also is the *place,* or space co-ordinate, of an event meaningful only if it enters our sensory experience while being subjected to measurement-in-principle, for example by means of meter sticks present on that occasion at the same time.

This was the kind of operationalist message which for most of his readers overshadowed all other philosophical aspects in Einstein's paper. This was the reason why it was embraced by the Vienna circle of neopositivists and by their related followers; [6] and this is why Besso, who had heard the message from

[5] Compare the similar analysis by R. S. Cohen: "[The phenomenalist program] suggested that nature was to be conceived as a set of disconnected atomic facts, that the flux of sensations can be analyzed into individual observation-protocols ... The phenomena with which science deals were assumed to be isolated sensations or single observations. The relations among the given phenomena were subjective matters of efficient but arbitrary ordering of the data; hypothetical entities and their relations were viewed as fictions or as shorthand; and the monadic character of atomic sensations was assumed *a priori* but made empirically plausible by a program of reductive definition of scientific concepts in terms of individual observation reports." (See: "Dialectical Materialsm and Carnap's Logical Empiricism," in *The Philosophy of Rudolf Carnap,* Ed. Paul A. Schilpp, La Salle, Illinois, 1963.)

[6] For example, see P. Frank, *Modern Science and its Philosophy,* Cambridge, 1949, pp. 61 to 89; V. Kraft, *The Vienna Circle,* New York, 1953; R. van Mises, *Ernst Mach und die Empiristische Wissenschaftsauffassung,* s'Gravenhage 1938 (reprinted from *Encyclop. der Einheitswissenschaft).* A warm welcome to relativity theory as "the victory over the metaphysics

Einstein before anyone else, exclaimed: "In the setting of Minkowski's space-time framework, it was now first possible to carry through the thought which the great mathematician, Bernhard Riemann, had grasped: 'The space-time framework itself is formed by the events in it'." [7]

To be sure, reading Einstein's paper with the wisdom of hindsight, we can find in it also very different trends that indicate the possibility that "reality" in the end is not going to be left identical with "events", that sensory experiences will in Einstein's later work not be regarded as the chief building blocks of the "world". Thus, the laws of physics themselves may be said to be also built into the event-world as the undergirding structure "governing" the pattern of events. We can see this even earlier, in one of Einstein's first letters now available in the Archives, dated 14 April 1901, to his friend, Marcel Grossmann. This is the period of Einstein's first scientific contributions, when he believed he had found a connection between the forces of attraction between molecules and Newtonian forces: "It is a wonderful feeling to recognize the unity of a complex of appearances which seem to direct sense experience to be separate things." Already here there is a tension between the priority ascribable to evident sense experience and intuited unity. But taking the early papers as a whole, and remembering the setting of the time, we realize that Einstein's philosophical pilgrimage *did* start on the historic ground of positivism.

3. *The Einstein-Mach Letters*

In the history of recent science, the relation between Einstein and Mach is a large and important topic. Here we can follow the four stages in the drama a little further: Einstein's early acceptance of the main features of Mach's doctrine; the Einstein-Mach correspondence and meeting; the revelation in 1921 of Mach's unexpected and vigorous attack on Einstein's relativity theory; and Einstein's own further development of a philosophy of knowledge in which he rejects many, but not all, of his earlier Machist tendencies.

So far, four letters have been found, addressed by Einstein to Mach. They are part of an exchange between 1909 and 1913, and they testify to Einstein's deeply felt attraction to Mach's idea at the time — a period in which the mighty Mach himself had apparently embraced the relativity theory publicly, for

of absolutes in the conceptions of space and time ..., a mighty impulse for the development of the philosophical point of view of our time" was extended by J. Petzoldt in the inaugural session of the *Gesellschaft für Positivistische Philosophie* in Berlin, 11 November 1912. (Reprinted in *Zeitschrift für Positivistische Philosophie, 1*, 1913, p. 4.)

[7] Letter of Besso to Einstein, 16 February 1939. Among many testimonies to the effect of Einstein on current and later positivistic philosophies of science, see P. W. Bridgman, "Einstein's Theory and the Operational Point of View", in *Albert Einstein, Philosopher-Scientist* (Ed. Paul A. Schilpp, Evanston, Illinois, 1949). Conversely, on the debt of Einstein's 1905 paper to Mach's philosophy, see P. Frank, *ibid.*, pp. 272–273, for example: "The definition of simultaneity in the special theory of relativity is based on Mach's requirement that every statement in physics has to state relations between observable quantities."

examples, in the second (1909) edition of his book on *Conservation of Energy:* "I subscribe, then, to the principle of relativity, which is also firmly upheld in my *Mechanics* and *Wärmelehre.*" [8]

Einstein writes from Bern in the first letter (of 9 August 1909), after thanking Mach for sending him the book on the law of conservation of energy: "I know, of course, your main publications very well, of which I most admire your book on Mechanics. You have had such a strong influence upon the epistemological conceptions of the younger generation of physicists that even your opponents today, such as Planck, undoubtedly would have been called Mach-followers by physicists of the kind that was typical a few decades ago."

It will be important for our analysis to remember that Planck, (Einstein's earliest patron in scientific circles, and who by 1913 succeeded in persuading his German colleagues to invite Einstein to the Kaiser-Wilhelm-Gesellschaft in Berlin), was indeed at that time the only scientifically prominent opponent of Mach, and had just written his famous attack, *Die Einheit des physikalischen Weltbildes* (1909). Far from accepting Mach's view that, as he put it, "nothing is real except the perceptions, and all natural science is ultimately an economic adaptation of our ideas to our perceptions", Planck held that a basic aim is "the finding of a *fixed* world independent of the variation of time and people" or, more generally, "the complete liberation of the physical picture from the individuality of the separate intellects". [9]

A reply from Mach, now lost, must have come quickly, because eight days later Einstein wrote again: "Bern, 17 August 1909. Your friendly letter gave me enormous pleasure ... I am very glad that you are pleased with the relativity theory ... Thanking you again cordially for your friendly letter, I remain, your student (indeed, *Ihr Sie verehrender Schüler).* A. Einstein."

The next letter was written just before or after Einstein's visit to Mach in the winter of 1911–1912, after the first progress toward the general relativity theory: "I can't quite understand how Planck has so little understanding for your efforts. His stand to my theory is also one of refusal. But I can't take it amiss; so far, that one single epistemological argument is the only thing which I can bring forward in favor of my theory." Einstein is referring delicately to the Mach Principle, which he had been putting at the center of the developing

[8] p. 95. For a brief analysis of Mach's expressions of both adherence and reservations with respect to the principle of relativity, see: H. Dingler, *Die Grundlagen der Machschen Philosophie*, Leipzig, 1924, pp. 73–86.

F. Herneck *(Phys. Blätter, 17,* 1961, p. 276) reports that P. Frank had the impression during a discussion with Ernst Mach around 1910 that Mach "was fully in accord with Einstein's special relativity theory and particularly with its philosophical basis" (letter to the author).

[9] *A Survey of Physical Theory,* New York, 1960, p. 24.

general relativity theory.[10] Mach responded by sending Einstein a copy of his book, *Analysis of Sensations.*

In the last of these letters to Mach (who was then 75 years old and for some years had been paralyzed), Einstein writes from Zurch on 25 June 1913:

> "Recently you have probably received my new publication on Relativity and Gravitation which I have at last finished after unending labor and painful doubt. [This must have been the *Entwurf*, with Marcel Grossmann.] Next year at the solar eclipse it will turn out whether the light rays are bent by the sun, in other words whether the basic and fundamental assumption of the equivalence of the acceleration of the reference frame and of the gravitational field really holds. If so, then your inspired investigations into the foundations of mechanics — despite Planck's unjust criticism — will receive a splendid confirmation. For it is a necessary consequence that inertia has its origin in a kind of mutual interaction of bodies, fully in the sense of your critique of Newton's bucket experiment." [11]

4. *The Paths Diverge*

While this correspondence stops here, Einstein's public and private avowals of his adherence to Mach's ideas continued for several years more. For example, there is his moving eulogy of Mach, published in 1916. In August 1918, Einstein writes to Besso quite sternly about an apparent lapse in Besso's positivistic epistemology; it is an interesting letter, worth citing in full:

> "28 August 1918.
>
> Dear Michele,
>
> In your last letter I find, on re-reading, something which makes me angry: that speculation has proved itself to be superior to empiricism. You are thinking here about the development of relativity theory. However, I find that this development teaches something else, that it is practically the opposite, namely that a theory which wishes to deserve trust must be built upon generalizable facts.
>
> Old examples: Chief postulates of thermodynamics [based] on impossibility of perpetuum mobile. Mechanics [based] on grasped [*ertasteten*] law of inertia. Kinetic gas theory [based] on equivalence of heat and mechanical energy (also historically). Special Relativity on constancy of light velocity and Maxwell's equation for the vacuum, which in turn rest on empirical foundations. Relativity with respect to constant translation is a *fact of experience.*
>
> General Relativity: *Equivalence of inertial and gravitational mass.* Never has a truly useful and deep-going theory really been found purely speculatively. The

[10] Later, Einstein found, of course, that this procedure did not work (see: *Ideas and Opinions*, New York, 1954, p. 286, and other publications). In a letter of 2 February 1954 to Felix Pirani, Einstein writes, "One shouldn't talk at all any longer of Mach's principle, in my opinion. It arose at a time when one thought that 'ponderable bodies' were the only physical reality and that in a theory all elements that are fully determined by them should be conscientiously avoided. I am quite aware of the fact that for a long time I, too, was influenced by this fixed idea."

[11] For a further analysis of the correspondence, see F. Herneck, *Forschungen und Fortschritte, 37*, 1963, p. 239; and H. Hönl, *Phys. Bl., 16*, 1960, p. 571.

nearest case is Maxwell's hypothesis concerning displacement current; there the problem was to do justice to the fact of light propagation ... With cordial greetings, your Albert." [Emphasis in original.]

Careful reading of this letter shows us that already here there is evidence of divergence between the conception of "fact" as understood by Einstein and "fact" as understood by Mach. The impossibility of the perpetuum mobile, the first law of Newton, the constancy of light velocity, the validity of Maxwell's equations, the equivalence of inert and gravitational mass — none of these would have been called "facts of experience" by Mach. Indeed, Mach might have insisted that — to use one of his favorite battle words — it is evidence of "dogmatism" not to regard all these conceptual constructs as continually in need of probing reexamination (cf. Mach, *Phys. Zs., 11,* 1910, p. 605). "For me, matter, time, and space are still *problems,* to which, incidentally, the physicists (Lorentz, Einstein, Minkowski) are also slowly approaching."

A similar divergence appears in a letter from Einstein to Ehrenfest (4 December 1919). Einstein writes: "I understand your difficulties with the development of relativity theory. They arise simply because you want to base the innovations of 1905 on epistemological grounds (non-existence of the stagnant ether) instead of empirical grounds (equivalence of all inertial systems with respect to light)." While Mach would have applauded Einstein's life-long suspicion of formal epistemological systems, how strange would he have found this use of the word "empirical" to characterize the hypothesis of equivalence of all inertial systems with respect to light! What we see here emerging slowly is Einstein's view, later frequently expressed explicitly, that the fundamental role played by experience in the making of fundamental physical theory is not through the "atom" of experience, not through the individual sensation or protocol sentence, but through *die gesamten Erfahrungstatsachen,* namely the *totality* of physical experience. [12]

But all along, apparently unknown to Einstein, a time bomb had been ticking away, even while he thought of Mach and himself close to each other's thoughts. In 1921, five years after Mach's death, there occurred the long-delayed publication of Mach's *Principles of Optics.* The Preface was dated July 1913 — a few days or, at most, weeks after Mach had received Einstein's last, enthusiastic letter. In a well-known passage, Mach and written:

"I am compelled, in what may be my last opportunity, to cancel my views [*Anschauungen*] of the relativity theory.

[12] Thus, as in the essay, "Space, Time and Gravitation," Einstein makes the distinction between constructive theories as against "theories of principle," of which he cites, as an example, the relativity theory. Such theories of principle, Einstein says, start with "empirically observed general properties of phenomena", and again the example of thermodynamics is cited (Space, Time and Gravitation, in *Out of My Later Years,* New York, 1950).

I gather from the publications which have reached, me, and especially from my correspondence, that I am gradually becoming regarded as the forerunner of relativity. I am able even now to picture approximately what new expositions and interpretations many of the ideas expressed in my book on Mechanics will receive in the future from this point of view.

It was expected that philosophers and physicists should carry on a crusade against me, for, as I have repeatedly observed, I was merely an unprejudiced rambler, endowed with original ideas, in varied fields of knowledge. I must, however, as assuredly disclaim to be forerunner of the relativists as I personally reject the atomistic doctrine of the present-day school or church. The reason why, and the extent to which, I reject [*ablehne*] the present-day relativity theory, which I find to be growing more and more dogmatical, together with the particular reasons which have led me to such a view — considerations based on the physiology of the senses, epistemological doubts, and above all the insight resulting from my experiments — must remain unchanged."

Certainly, Einstein was deeply disappointed by Mach's explicit dismissal of the relativity theory (except possibly for its "fruitfulness in pure mathematics"). During a talk on 6 April 1922 in Paris, in a discussion with Emile Meyerson, Einstein allowed that Mach was "un bon mécanicien," but "undéplorable philosophe."

We can well understand that at heart Mach's rejection was very painful to Einstein, the more so as it was somehow Einstein's tragic fate to have the contribution he most cared about be rejected by the very men whose approval and understanding he would have most gladly had — a situation not unknown in the history of science. In addition to Mach, the list includes these four: Poincaré (who, to his death in 1912, only once deigned to mention Einstein's name in print, and then only to register his dissent); Lorentz (who gave Einstein personally every possible encouragement — short of fully accepting this theory of relativity for himself); Planck (whose support of the special theory of relativity was unstinting but who resisted Einsteins ideas on general relativity, not to speak of the early quantum theory of radiation); and Michelson, who to the end of his days did not believe in relativity theory, and even once said to Einstein that he was sorry that his own work may have helped to start this "monster". [13]

However, by and by, Einstein's generosity took again the upper hand and resulted in many further personal testimonies to Mach's earlier influence. One example has been given — the note in the *Neue Freie Presse* of 1926. Another typical example is a letter of 18 September 1930 to Armin Weiner:

"I did not have a particularly important exchange of letters with Mach. However, Mach did have a considerable influence upon my development through his writings. Whether or to what extent my life's work was influenced thereby is

[13] R. S. Shankland, *American Journal of Physics, 31,* 1963, p. 56.

impossible for me to find out. Mach occupied himself in his last years with the relativity theory, and in a preface to a late edition of one of his works even spoke out in rather sharp refusal against the relativity theory. However, there can be no doubt that this was a consequence of a lessening ability to take up [new ideas] owing to his age, for the whole direction of thought of this theory conforms with Mach's, so that Mach quite rightly is considered as a forerunner of general relativity theory." [14]

A revealing summary by Einstein himself was provided in his letter of 8 January 1948 to Besso:

"As far as Mach is concerned, I wish to differentiate between Mach's influence in general and his influence on me... Particularly in the *Mechanics* and the *Wärmelehre* he tried to show how conceptions arose out of experience. He took convincingly the position that these conceptions, even the most fundamental ones, obtained their warrant only out of empirical knowledge, that they are in no way logically necessary... I see his weakness in this, that he more or less believed science to consist in a mere ordering of empirical material; that is to say, he did not recognize the freely constructive element in formation of concepts. In a way he thought that theories arise through *discoveries* and not through *inventions*. He even went so far that he regarded "sensations" not only as material which has to be investigated, but as it were, as the building blocks of the real world; thereby, he believed, he could overcome the difference between psychology and physics. If he had drawn the full consequences, he would have and had to reject not only atomism but also the idea of a physical reality.

Now, as far as Mach's influence on my own development is concerned, it certainly was great. I remember very well that you drew my attention to his *Mechanics* and *Wärmelehre* during my first years of study, and that both books made a great impression on me. The extent to which they influenced my own work is, to say the truth, not clear to me. As far as I am conscious of it, the immediate influence of Hume on me was greater... But, as I said, I am not able to analyze that which lies anchored in unconscious thought. It is interesting, by the way, that Mach rejected the special relativity theory passionately (he did not live to see the general relativity theory). The theory was, for him, inadmissibly speculative. He did not know that this speculative character belongs also to Newton's mechanics, and to every theory which thought is capable of. There

[14] I thank Colonel Dibner for making a copy of the letter available to me from the Archives of the Burndy Library in Norwalk, Connecticut. Among other hitherto unpublished letters in which Einstein indicated his indebtedness to Mach, we may cite one to A. Lampa (9 December 1935): "You speak about Mach as about a man who has gone into oblivion. I cannot believe that this corresponds to the facts since the philosophical orientation of the physicists today is rather close to that of Mach, a circumstance which rests not a little on the influence of Mach's writings." Moreover, practically everyone else shared Einstein's explicitly expressed opinion of the debt of relativity to Mach; thus H. Reichenbach wrote in 1921 (*Logos*, *10*, p. 331): "Einstein's theory signifies the accomplishment of Mach's program." A similar view is still widely expressed recently, for example by H. Schardin, the Head of the Ernst-Mach-Institut of Freiburg i. Br., in his preface to the useful book by K. D. Heller, *Ernst Mach*, Vienna-New York, 1964.

exists only a gradual difference between theories, insofar as the chains of thought from fundamental concepts to empirically verifiable conclusions are of different lengths and complications."

There are additional resources, both published and unpublished, on the detailed aspects of the relation between Einstein and Mach, which, for lack of space, cannot be summarized here.

5. *Antipositivistic Component of Einstein's Work*

Mach's own harsh words in the 1913 Preface leave a tantalizing mystery. For example, Ludwig Mach's act of destroying many of his father's papers has so far made it impossible to find out more about the "experiments" (possibly on the constancy of the velocity of light) at which Mach hinted. Since 1921, many speculations have been offered to explain Mach's action [15]; they can all be simultaneously true, although they all leave something to be desired.

And yet, I believe it is not so difficult to reconstruct the main reasons why Mach ended up rejecting the relativity theory (leaving aside speculations on the possible contributions to this rift owing to the difference between Einstein and Mach on atomism). To put it very simply, Mach had more and more clearly recognized, years before Einstein did so himself, that Einstein had left behind him the confines of Machist empirio-critical phenology.

The list of evidences is long; here only a few examples can be given. First, the initial 1905 relativity paper itself: what had made it really work was not merely the empiricist-operationist component, but the initial, courageous postulation of the two thematic hypotheses (concerning the constancy of light velocity and the relativity in all branches of physics), two postulates for which there was no direct confirmation at all. For a long time, Einstein did not draw attention to this feature. In a lecture at King's College, London, in 1921, just before the publication of Mach's posthumous attack, Einstein still was protesting that the origin of relativity theory lay in the facts of direct experience:

"I am anxious to draw attention to the fact that this theory is not speculative in origin; it owes its invention entirely to the desire to make physical theory fit observed fact as well as possible. We have here no revolutionary act but the natural continuation of a line that can be traced through centuries. The abandonment of certain notions connected with space, time, and motion, hitherto treated as fundamentals, must not be reagraded as arbitrary, but only as conditioned by observed facts." [16]

[15] E. g. by Einstein himself, J. Petzoldt, and H. Dingler (cf. Dingler, *op. cit.*, pp. 84–86). F. Herneck adds the significant report that according to a letter from P. Frank, Mach was personally influenced by Dingler "who was an embittered opponent of Einstein" *(op. cit.*, p. 276).

[16] "*On the Theory of Relativity*", in *Mein Weltbild*, Amsterdam, 1934; republished in *Ideas and Opinions*, Crown Publishers, New York, 1954, p. 246.

But by June 1933, when Einstein returned to England to give the Herbert Spencer Lecture at Oxford, the much more complex, sophisticated epistemology which in fact was inherent in his work from the outset had begun to assert itself. He opened this lecture with the significant sentence, "If you want to find out anything from the theoretical physicists about the methods they use, I advise you to stick closely to one principle: don't listen to their words, fix your attention on their deeds". This lecture divides the tasks of experience and reason in a very different way from that advocated in his earlier lecture: "We are concerned with the eternal antithesis between the two inseparable components of our knowledge, the empirical and the rational, in our department... The structure of the system is the work of reason; the empirical contents and their mutual relations must find their representation in the conclusions of the theory. In the possibility of such a representation lies the sole value and justification of the whole system, and especially the concepts and fundamental principles which underlie it. Apart from that, these latter are free inventions of the human intellect, which cannot be justified either by the nature of that intellect or in any other fashion *a priori*." And in summary of this passage he draws attention to the "purely fictitious character of the fundamentals of scientific theory". It is precisely this penetrating insight which Mach must have smelled out much earlier, and dismissed as "dogmatism".

Indeed Einstein, in his Spencer Lecture, addresses himself to the old view that "the fundamental concepts and postulates of physics were not in the logical sense inventions of the human mind but could be deduced from experience by 'abstraction' — that is to say, by logical means. A clear recognition of the erroneousness of this notion really only came with the general theory of relativity." Einstein ends this discussion with the clear enunciation of his credo: "Nature is the realization of the simplest conceivable mathematical ideas. I am convinced that we can discover by means of purely mathematical constructions the concepts and the laws connecting with each other, which furnished the key to the understanding of natural phenomena. Experience may suggest the appropriate mathematical concepts, but they most certainly cannot be deduced from it. Experience remains, of course, the sole criterion of physical utility of a mathematical construction. But the creative principle resides in mathematics. In a certain sense, therefore, I hold it true that pure thought can grasp reality, as the ancients dreamed." [17]

[17] Quotations from "On the Method of Theoretical Physics", in *Mein Weltbild*, 1934; reprinted in *Ideas and Opinions*, pp. 270–276. There are a number of later lectures and essays in which the same point is made, particularly the lecture, "Physics and Reality", which explicitly states that Mach's theory of knowledge "on account of the relative closeness of the concepts used to experience" did not suffice, and that one must go "beyond this phenomenological point of view" to achieve a theory whose basis is "further removed from direct experiment but more uniform in character".

Even as Einstein's views developed, so did those of many of the philosophers of science who also had earlier started from a more strict Machist position. This growing modification

Einstein himself stressed later the key role of thematic rather than phenomenic elements in his earliest work. Thus he wrote in a famous passage in his *Autobiographical Notes* (1949):

> "Reflections of this type made it clear to me as long ago as shortly after 1900, i.e., is shortly after Planck's trail-blazing work, that neither mechanics nor thermodynamics could (except in limiting cases) claim exact validity. By and by I despaired of the possibility *[Nach und nach verzweifelte ich an der Möglichkeit]* of discovering the true laws by means of constructive efforts *based on known facts.* The longer and the more despairingly I tried *[Je länger und verzweifelter ich mich bemühte],* the more I came to the conviction that only the discovery of *a universal formal principle* could lead us to assured results" (p. 53; emphasis added. G. H).

A second example to show the evidence of a gradual disengagement from a Machist position is a very early one: it comes from Einstein's article on relativity in the 1907 *Jahrbuch der Radioaktivität und Elektronik (4,* No. 4), where Einstein discusses the experiments by W. Kaufmann, and in particular Kaufmann's paper in the *Annalen der Physik* (*19,* 1906), "Concerning the Constitution of the Electron". This paper was the first one in the *Annalen der Physik* to mention Einstein's work on the relativity theory, and coming from the eminent experimental physicist Kaufmann, it was most significant that this first discussion was intended to be a categorical experimental disproof of Einstein's theory. Kaufmann announced:

> "I anticipate right here the general result of the measurements to be described in following: *the measurement results are not compatible with the Lorentz-Einsteinian fundamental assumption.*" [18]

Not until 1916, through the work of Guye and Lavanchy, was the inadequacy of Kaufmann's equipment fully discovered.

Now in his discussion of 1907, Einstein acknowledges that there is a systematic small difference between Kaufmann's results and his own predictions. Kaufmann's calculations seem free of error, but "whether there is an unsuspected systematic error or whether the foundations of relativity theory do not correspond with the facts, one will be able to decide with certainty only if a great variety of observational material is at hand." But despite this prophetic remark, Einstein does not rest his case on it. On the contrary, he has a very different, and what for his time must have been a very daring, point to make: he acknowledges that the theories of electron motion, given earlier by Abraham

of original position, partly owing to "the growing understanding of the general theory of relativity", has been chronicled by P. Frank (e. g., in "*Einstein, Mach, and Logical Positivism*", *in* Schilpp, *op. cit.).*

[18] W. Kaufmann, *op. cit.,* p. 495; emphasis in original.

and by Bucherer, do give predictions considerably closer to the experimental results of Kaufmann, but Einstein refuses to let the "facts" decide the matter: "In my opinion both theories have a rather small probability because their fundamental assumptions concerning the mass of moving electrons are not explainable in terms of theoretical systems which embrace a greater complex of phenomena" (p. 439).

This is the characteristic position — the crucial difference between Einstein and those who would make the correspondence with experimental fact the chief deciding factor for or against a theory: even though the "experimental facts" at that time very clearly seem to favor the theory of his opponents rather than his own, he finds the *ad hoc* character of their theories more objectionable than the apparent disagreement between his theory and the "facts".

So already in this 1907 article (which, incidentally, Einstein mentions in his postcard of 17 August 1909 to Ernst Mach, with a remark regretting that he has no more reprints for distribution) we have explicit evidence of a gradual hardening of Einstein against the epistemological priority of experiment, not to speak of sensory experience. More and more clearly, Einstein puts the consistency of a simple and convincing theory or of a thematic conception higher in importance than the latest news from the laboratory — and again and again, he turns out to be right.

Thus, a few months after Einstein had written, in his fourth letter to Mach, that the solar experiment will decide "whether the basic and fundamental assumption of the equivalence of the acceleration of the reference frame and of the gravitational field really holds", Einstein writes to Besso in a very different vein (in March 1914), before the first, ill-fated eclipse expedition to test the conclusions of the preliminary version of the general relativity theory: "Now I am fully satisfied, and I do not doubt anymore the correctness of the whole system, may the observation of the eclipse succeed or not. The sense of the thing [*Vernunft der Sache*] is too evident." And later, commenting on the fact that there remains up to ten per cent discrepancy between the measurement of the deviation of light by the sun's field and the calculated effect based on the general relativity theory: "For the expert, this thing is not particularly important because the main significance of the theory does not lie in the verification of little effects, but rather in the great simplification of the theoretical basis of physics as a whole." [19] Or again, in his *Notes on the Origin of the General Theory of Relativity* [20], Einstein reports that he "was in the highest degree amazed" at the existence of the equivalence between inertial and gravitational mass, but that he "had no serious doubts about its strict validity, even without knowing the results of the admirable experiment of Eötvös".

[19] Seelig, *op. cit.*, p. 195.

[20] *Mein Weltbild*, Amsterdam, 1934; reprinted in *Ideas and Opinions*, New York, 1954, pp. 285–290.

The same important point is made again in a revealing account given by Einstein's student, Ilse Rosenthal-Schneider. In a manuscript "Reminiscences of Conversation with Einstein," dated 23 July 1957, she reports:

> "Once when I was with Einstein in order to read with him a work that contained many objections against his theory ... he suddenly interrupted the discussion of the book, reached for a telegram that was lying on the window sill and handed it to me with the words, 'Here, this will perhaps interest you'. It was Eddington's cable with the results of measurement of the eclipse expedition [1919]. When I was giving expression to my joy that the results coincided with his calculations, he said quite unmoved, 'But I knew that the theory is correct', and when I responded, but what if there had been no confirmation of his prediction, he countered: 'Then I would have been sorry for the dear Lord — the theory *is* correct.'"

6. *Minkowski's "World" vs. the World of Sensations*

The third major point at which Mach, if not Einstein himself, must have seen that their paths were diverging is the development of relativity theory into the theory of the geometry of the four-dimensional space-time continuum, begun by Minkowski in 1908. We have several indications that Mach was concerned about the introduction of four-dimensional geometry into physics; for example, according to F. Herneck,[21] Ernst Mach specially invited the physicist and philosopher, Phillip Frank, to visit him, specifically "in order to find out more about the relativity theory, above all about the use of four-dimensional geometry". Perhaps as a result, P. Frank published the review article on "The Principle of Relativity and the Representation of Physical Phenomena in Four-Dimensional Space".[22] It is an attempt, addressed to readers "who do not master modern mathematical methods", to show that Minkowski's essay brings out the "empirical facts far more clearly by the use of four-dimensional world lines". The essay ends with the conclusion: "In this four-dimensional world the facts of experience can be presented more adequately than in three-dimensional space, where always only an arbitrary and one-sided projection is pictured."

Following Minkowski's own paper, Frank's treatment can make it still appear in most respects that the time dimension is equivalent to the space dimensions. Thereby one could think of Minkowski's treatment as basing itself not only on a functional operational interconnection of space and time — which is fully in accord with Mach's own views — but also on the primacy of ordinary space and time in the relativistic description of phenomena. Initially, Mach seems to have been hospitable to Minkowski's presentation; thus Mach wrote in the 1909 edition of *Conservation of Energy:* "Space and time are not

[21] *Physikalische Blätter, 15*, 1959, p. 565.

[22] *„Das Relativitätsprinzip und die Darstellung der physikalischen Erscheinungen im vierdimensionalen Raum,"* Z. *für Phys. Chemie*, 1910, pp. 466–495.

here conceived as independent entities, but as forms of the dependence of the phenomena on one another," followed by the citation of Minkowski's just-published essay, *Space and Time*. But a few lines earlier, Mach had written: "Spaces of many dimensions seem to me not so essential for physics. I would only uphold them if things of thought like atoms are maintained to be indispensable, and if, then, also the freedom of working hypothesis is upheld."

It was correctly pointed out by C. B. Weinberg [23] that Mach may eventually have had two sources of suspicion against the Minkowskian form of relativity theory. As was noted above, Mach regarded the fundamental notions of mechanics as problems to be continually discussed with maximum openness within the frame of empiricism, rather than as questions that can be solved and settled — as the relativists, seemingly dogmatic and sure of themselves, were in his opinion more and more inclined to do. In addition, Mach held that the questions of physics were to be studied in a broader setting, encompassing biology and psychophysiology. Thus Mach wrote: "Also, physics is not the entire world; biology is there too, and belongs essentially to the world picture". [24]

But there is also a third reason for Mach's eventual antagonism against such conceptions — except if they are applied to "mere things of thought like atoms and molecules, which by their very nature can never be made the objects of sensuous contemplations". [25] For if one takes Minkowski's essay, and what followed from it, seriously (for example the abandonment of space and time separately, with only "a kind of union of the two" having identity), one must recognize that this abandonment of the conceptions of experiential space and experiential time is an attack on the very roots of phenomenalism or sensations-physics, namely on the meaning of actual measurements. If identity, meaning, or "reality" lie in the four-dimensional space-time interval *ds*, one deals now with a quantity which is hardly *denkökonomisch* nor one that preserves the primacy of measurements in "real" space and time. In his exuberant article, Minkowski had announced that "three-dimensional geometry becomes a chapter in four-dimensional physics... Space and time are to fade away into shadows, and only *eine Welt an sich* will subsist." And in this world the crucial innovation is the conception of the *zeitartigen Vektorelement*

$$\mathrm{ds}\left(=\frac{1}{c}\sqrt{c^2\,dt^2 - dx^2 - dy^2 - dz^2}\right)$$

with imaginary components.

To Mach the very word "element" had a crucial and very different meaning: the elements were nothing less than the sensations and complexes of sensations

[23] *Mach's Empirio-Pragmatism in Physical Science, thesis*, Columbia University, 1937.

[24] *Scientia*, 7, 1910, p. 225.

[25] *Space and Geometry*, 1906, p. 138. Mach's attempts to speculate on the use of n-dimensional spaces for representing the configuration of such "mere things of thought" — the derogatory phrase also applied to absolute space and absolute motion in Newton — are found in his first major book, *Conservation of Energy* (1872).

of which the world consists and which completely define the world. What relativity theory was now clearly doing, as seen from this point of view, was to move the ground of basic, elemental truths from the plane of direct, naked experience in ordinary space and in time to a mathematicized model of the world in a union of space and time that is not directly accessible to sensation — and in this respect, reminiscent of absolute space and time, which Mach had called "metaphysical monsters".

To see this point clearly, it will be useful to consider a brief summary of an essential part of Mach's philosophy as presented by his sympathetic follower, Moritz Schlick: [26]

"Mach was a physicist, physiologist, and also psychologist, and his philosophy, as just mentioned, arose from the wish to find a principal point of view to which he could hew in any research, one which he would not have to change when going from the field of physics to that of physiology or psychology. Such a firm point of view he reached by going back to that which is given before all scientific research: namely, the world of sensations ... Since all our testimonies concerning the so-called external world rely only on sensations, Mach held that we can and must take these sensations and complexes of sensations to be the sole content *[Gegenstände]* of those testimonies, and, therefore, that there is no need to assume in addition an unknown reality hidden behind the sensations. With that, the existence *der Dinge an sich* is removed as an unjustified and unnecessary assumption. A body, a physical object, is nothing else than a complex, a more or less firm [we would say 'invariant'] pattern of sensations, that is of colors, sounds, heat and pressure, etc. There exists in this world nothing whatever other than sensations and their connections. In place of the word 'sensations', Mach liked to use rather the more neutral word 'elements'."

"[As is particularly clear in Mach's book, *Erkenntnis und Irrtum*,] scientific knowledge of the world consists, according to Mach, in nothing else than the simplest possible description of the connections between the elements, and it has as its only aim the intellectual mastery of those facts by means of the least possible effort of thought. [27] This aim is reached by means of a more and more complete 'accommodation of thoughts to the facts and the accommodation of the thoughts to one another'. This is the formulation by Mach of his famous 'principle of the economy of thought'."

Here, then, is an issue which, from the beginning, separated Einstein and Mach, even before they realized it. To the latter the fundamental task of science was economic and descriptive; to the former, it was creative and intuitive. Mach had once written: "If all the individual facts — all the individual phenomena, knowledge of which we desire — were immediately acces-

[26] In the essay, *Ernst Mach, Der Philosoph;* also published in the special supplement on Ernst Mach in the *Neue Freie Presse*, Vienna (12 June 1926).

[27] It is interesting to note that this theme appears already in Mach's first popular essay, *Die Gestalten der Flüssigkeiten*, 1868.

sible to us, science would never have arisen" (*Conservation of Energy*, p. 54). To this, with the forthrightness caused perhaps by his recent discovery of Mach's opposition, Einstein countered during his lecture in Paris of 6 April 1912: "Mach's system studies the existing relations between data of experience; for Mach, science is the totality of these relations. That point of view is wrong and, in fact, what Mach has done is to make a catalog, not a system." [28]

We realize that we are witnessing here an old conflict, and one that has continued throughout the development of the sciences. Mach's phenomenalistic positivism, with its almost Baconian invocation of the primacy of sensory experience, brandished an undeniable and irresistible weapon for the critical reevaluation of classical physics; and in this it seems to hark back to an ancient position that looked upon sensuous appearances as the beginning and end of scientific achievement. Once can read Galileo in this light, when he urges the primary need of *description* for the fall of bodies, leaving "the causes" to be found out later. So one can understand (or rather, misunderstand) Newton, with his too well-remembered remark, "I feign no hypotheses". [29] Kirchhoff is in this tradition — Boltzmann wrote of him in 1888: "The aim is not to produce bold hypotheses as to the essence of matter, or to explain the movement of a body from that of molecules, but to present equations which, free from hypotheses, are as far as possible true and quantitatively correct correspondents of the phenomenal world, careless of the essence of things and forces. In his book on mechanics, Kirchhoff will ban all metaphysical concepts, such as forces, the cause of a motion; he seeks only the equations which correspond so far as possible to observed motions". [30] And so could, and did, Einstein himself understand the Machist component of his own early work.

But phenomenalistic positivsm in science has always been victorious only up to a very definite limit. It was the necessary sword for destroying old error, but it made a very inadequate plowshare for cultivating a new harvest. I find it exceedingly significant that Einstein saw this during the transition phase of partial disengagement — a disengagement from Machist philosophy which, to be sure, never became quite complete. In the spring of 1917 Einstein wrote to Besso and mentioned a manuscript which Friedrich Adler had sent him. Einstein commented, "He rides Mach's poor horse to exhaustion". To this Besso, almost always the loyal Machist, responds on 5 May 1917: "As to Mach's little horse, we should not insult it; did it not make possible the infernal journey through the relativities? And who knows — in the case of the nasty quanta, it

[28] Reported in "Einstein and the Philosophies of Kant and Mach", *Nature, 112*, August 1923, p. 253.

[29] That Einstein did not so misunderstand Newton can be illustrated, for example, in a comment reported by C. B. Weinberg: "Dr. Einstein further maintained that Mach, as well as Newton, tacitly employs hypotheses — not recognizing their non-empirical foundations" (*op. cit.*, p. 55). For an example of Mach's tacit presuppositions, see H. Dingler, *op. cit.*, pp. 69–71.

[30] Cited by R. S. Cohen, *op. cit.*, p. 109.

may also carry Don Quixote de la Einsta through it all!" Einstein's answer (13 May 1917) is revealing: "I do not inveigh against Mach's little horse; but you know what I think about it. It cannot give birth to anything living, it can only exterminate harmful vermin."

7. Toward a Rationalistic Realism

Thus, when the time had come to stop the re-evaluation and to begin the reconstruction, Mach's message by itself was insufficient. (In this sense, Mach, like many masters, rightly feared his disciple.) Indeed, what Einstein had done was first to adopt Mach's doctrine (as in a crucial part of the 1905 paper), and then to turn it upside down — minimizing rather than maximizing the role of actual details of experience, both at the beginning and at the end of scientific theory, and opting for a creative rationalism that almost inevitably would lead him to the conception of an objective, "real" world behind the phenomena to which our senses are exposed. (This is, of course, the same choice which Galileo, Newton, and Planck made.)

By 1931, in the essay "Maxwell's Influence on the Evolution of the Idea of Physical Reality", Einstein could start with the words: "The belief in an external world independent of the perceiving subject is the basis of all natural science". Again and again, in the period beginning with his work on the general relativity theory, Einstein insisted that between experience and reason, as well as between the world of sensory perception and the objective world, there are logically unbridgeable chasms. The efficacy of reason to grasp reality was characterized later by Einstein by the word "miraculous". Even the very terminology in these statements would have been anathema to Mach.

At this point we may well ask when and under what circumstances Einstein himself slowly became aware of his deep-going epistemological change, a change that, to put it very briefly, consisted in realizing and accentuating more and more clearly the rationalistic element in his work, and correspondingly realizing and accentuating less and less the positivistic elements (and not, by any means, any dramatic reversal against one "pure" position and acceptance of another that is diametrically opposed). Here again, we may turn for illumination to one of the hitherto-unpublished letters, one written to his old friend, C. Lanczos, on 24 January 1938:

> "Coming from sceptical empiricism of somewhat the kind of Mach's, I was made, by the problem of gravitation, into a believing rationalist, that is, one who seeks the only trustworthy source of truth in mathematical simplicity. The logically simple does not, of course, have to be physically true, but the physically true is logically simple, that is, it has unity at the foundation."

Indeed, all other evidences point to the conclusion that Einstein's work on general relativity theory was crucial in his epistemological development. As he wrote later (*Physics and Reality*, 1936): "The first aim of the general theory of relativity was the preliminary version which, while not meeting the requirements for constituting a closed system, could be connected in as simple a manner as possible with 'directly observed facts'." But the aim could not be achieved. In "Notes on the Origin of the General Relativity Theory,"[31] Einstein reported: "I soon saw that the inclusion of non-linear transformation, as the principle of equivalence demanded, was inevitably fatal to the simple physical interpretation of the coordinates, — that is, that it could no longer be required that coordinate differences should signify direct results of measurement with ideal scales or clocks. I was much bothered by this piece of knowledge..." — just as Mach must have been. "The solution of the above mentioned dilemma [from 1912 on] was therefore as follows: a physical significance attaches not to the differentials of the coordinates, but only to the Riemannian metric corresponding to them."[32]

It was the final consequence of the Minkowskian four-space representation. And it was the choice that had to be made — against fidelity to a chaos of operational experience, and in favor of fidelity to the ancient hope for a unity at the base of physical theory.

I am not touching in this essay on the effect of Quantum Mechanics on Einstein's epistemological development, a chief reason being that while from his "heuristic" announcement of the value of quantum theory in 1905 Einstein remained consistently skeptical about the "reality" of the quantum theory of radiation, this opinion only added to the growing realism stemming from his work on general relativity theory.

Four quotations from Einstein's letters, widely separated in time, will help to illustrate this point in a concise manner.

Letter to Hopf (probably of December 1911): "The quanta do what they should, but they do not exist — just as the resting luminiferous ether."

Letter to Max Born (24 April 1924): "Bohr's opinion of radiation interests me very much. But I don't want to let myself be driven to a renunciation of strict causality before there has been a much stronger resistance against it than up to now... It is true, my attempts to give the quanta palpable shape have failed again and again, but I'm not going to give up hope for a long time yet."

Letter to Max Born (7 September 1944): "In our scientific expectations we have become antipodes. You believe in the dice-playing God, and I in perfect rules of law in a world of something objectively existing, which I try to catch in a wildly speculative way."

Letter to Max Born (2 December 1947): "The theory is incompatible with the principle that physics is to present a reality in space and time."[33] Similarly, in 1940 Einstein wrote he "cannot believe that we must abandon, actually and forever, the idea of direct representation of physical reality in

[31] *Op. cit.*, p. 288.

[32] *Op. cit.*, p. 289.

[33] The three excerpts from Born letters were published by Max Born in *Universitas, 8*, 1965, pp. 33–44.

space and time."[34] Although Einstein had taken this step in General Relativity, he could not do so in Quantum Mechanics.

Enough has been written in other places to show the deepgoing connections which existed between Einstein's scientific rationalism on the one hand, and his philosophical realism and belief in a form of cosmic religion on the other. Max Born summarized it in one sentence: "He believed in the power of reason to guess the laws according to which God has built the world."[35] Perhaps the best expression of this position by Einstein himself is to be found in his much-neglected essay, *Über den gegenwärtigen Stand der Feld-Theorie.*[36]

Physical theory has "two ardent desires": to gather up as far as possible all pertinent phenomena and their connections, and to help us "not only to know *how* Nature is and *how* her transactions are carried through, but also to reach as far as possible the perhaps utopian and seemingly arrogant aim of knowing why Nature is *thus and not otherwise.* Here lies the highest satisfaction of a scientific person." [On making deductions from a "fundamental hypothesis" such as that of the molecular-kinetic theory] "one experiences so to speak that God Himself could not have arranged those connections [for example between pressure, volume, and temperature] in any other way than that which factually exists, any more than it would be in His power to make the number 4 into a prime number. This is the promethean element of the scientific experience... Here has always been for me the particular magic of scientific considerations; this is, so to say, the religious basis of scientific effort."

This fervor is indeed far from the kind of analyses which Einstein had made only a few years earlier — doubly far from the ascetism of his first philosophic mentor Mach, who had written in his day book: "Colors, space, tones, etc. These are the only realities. Others do not exist."[37] And far closer to the rational realism of his first scientific mentor, Planck, who wrote: "The disjointed data of experience can never furnish a veritable science without the intelligent interference of a spirit actuated by faith... We have a right to feel secure in surrendering to our belief in a philosophy of the world based upon a faith in the rational ordering of this world."[38] Indeed, we note immediately the profound philosophical kinship of Einstein's position with seventeenth century natural philosophers — for example, with Johannes Kepler, who in the Preface of the *Mysterium Cosmographicum* announced he wanted to find out concerning the number, positions, and motions of the planets, "why they as they are, and not otherwise", and who wrote to Herwart in April 1599 that

[34] *Ideas and Opinions*, p. 334.

[35] Max Born, "Physics and Relativity," in *Physics in My Generation*, London, 1956, p. 205.

[36] Published in the *Festschrift für Aurel Stodola*, Orell Füssli Verlag, Zurich and Leipzig, 1929, pp. 126–132.

[37] H. Dingler, *op. cit.*, p. 98.

[38] *The Philosophy of Physics*, New York, 1936, pp. 122 and 125.

with regard to numbers and quantity, "our knowledge is of the same kind as God's, at least insofar as we can understand something of it in this mortal life."

Needless to say, Einstein's friends from earlier, more positivistically inclined days, sometimes had to be informed of the change in a direct manner. For example, Einstein wrote to Moritz Schlick on 28 November 1930: "In general your presentation fails to correspond to my conceptual style insofar as I find your whole orientation so to speak too positivistic... I tell you straight out: physics is the attempt at the conceptual construction of a model of the *real world* and of its lawful structure. To be sure, it [physics] must present exactly the empirical relations between those sense experiences to which we are open; but only *in this way* is it chained to them... In short, I suffer under the (unsharp) separation of experience-reality and reality-of-Being... You will be astonished about the 'metaphysicist' Einstein. But every four- and two-legged animal is de facto in this sense metaphysicist" (emphasis in original).

Similarly, P. Frank, Einstein's early associate and later his biographer, reports that the realization of Einstein's true state of thought reached him in a most embarassing way, at a congress of German physicists in Prague in 1929, just as Frank was delivering "an address in which I attacked the metaphysical position of the German physicists and defended the positivistic ideas of Mach". The very next speaker showed Frank that he had been mistaken still to associate Einstein's views with those of Mach and himself. "He added that Einstein was entirely in accord with Planck's view that physical laws describe a reality in space and time that is independent of ourselves."

"At that time", Frank adds, "this presentation of Einstein's views took me very much by surprise." [39] But of course, in retrospect it is much easier to see the beginning of this change, one which Mach, Planck's main philosophical target, himself had seen in 1913, before anyone else. The events in the years around 1930 served merely as the occasions for private and public announcement of a long development. In particular, Einstein himself now realized how close he had moved to Planck, from whom he had explicitly dissociated himself in three of the letters to Mach. For example, in the Einstein Archives there is a handwritten draft, written on or just before 17 April 1931, and intended as introduction to Planck's hard-hitting article "Positivism and the Real External World". [40] In lauding Planck's article, Einstein concludes: "I presume I may add that both Planck's conception of the logical state of affairs as well as his subjective expectation conccerning the later development of our science corresponds entirely with my own understanding".

In his above-mentioned essay, Planck made a clear exposition of his views both in physics and in philosophy more generally. Thus he wrote: "Hitherto,

[39] P. Frank, *Einstein, His Life and Times*, New York, 1947, p. 215.

[40] In *International Forum, 1*, 1931, Nos. 1 and 2. Einstein sent his introduction to the Editor of the journal on 17 April 1931, but it appears to have come too late for inclusion.

the principle of causality was universally accepted; but now even that has been thrown overboard. That such an extraordinary fact should have taken place in the realm of physical science is widely said to be a symptom of the all-round unreliability of human knowledge. As a physicist I may be permitted to put forward some views of my own on the situation in which physical science finds iteself now, when confronted with the questions and problems to which I have referred. Perhaps what I shall have to say may throw some light also on other fields of human activity which the cloud of skepticism has darkened" (p. 12).

"The essential point of the positivist theory is that there is no other source of knowledge except the straight and short way of perception through the senses. Positivism always holds strictly to that. Now, the two sentences: 1) *There is a real outer world which exists independently of our act of knowing,* and 2) *The real other world is not directly knowable,* form together the cardinal hinge on which the whole structure of physical science turns. And yet there is a certain degree of contradiction between those two sentences. This fact discloses the presence of the irrational, or mystic, element which adheres to physical science as to every other branch of human knowledge. The effect of this is that a science is never in a position completely and exhaustively to solve the problem it has to face. We must accept that as a hard and fast irrefutable fact, and this fact cannot be removed by a theory which restricts the scope of science at its very start. Therefore, we see the task of science arising before as an incessant measuring struggle toward a goal which will never be reached, because of its very nature it is unreachable. It is of a metaphysical character and, as such, is always again and again beyond our achievement" (pp. 15–16).

In an essay in honor of Bertrand Russell, Einstein warns that the "fateful 'fear of metaphysics'... has come to be a malady of contemporary empiricistic philosophizing".[41]

And in their numerous letters the two old friends, Einstein and Besso, to the very end touchingly and patiently try to explain their positions and perhaps change the other's. Thus on 28 February 1952 Besso once more presents a way of making Mach acceptable to Einstein. The latter, in answering on 20 March 1952, once more responds that the facts cannot lead to a deductive theory, and at most can set the stage "for intuiting a general principle" as the basis of a deductive theory. And a little later, Besso had to be gently scolded (letter of 13 July 1952): "It appears that you do not take the four-dimensionality of reality, but that instead you take the present to be the only reality. What you call 'world' is in physical terminology 'spacelike section' for which the relativity theory — already the special theory — denies objective reality."

[41] *The Philosophy of Bertrand Russell,* P. Schilpp, ed., Evanston, Illinois, 1944, p. 289.

In the end, Einstein came back full circle to a view which many (and perhaps he himself) thought he had eliminated from physics in the basic 1905 paper on relativity theory:[42] there exists an external, objective, physical reality which we may hope to grasp — not directly empirically or logically or with fullest certainty, but at least by an intuitive leap, one that is only guided by experience with the totality of sensible "facts"; events happen in a "real world", of which the space-time world of sensory experience, and even the world of multidimensional continua, are useful metaphors, but no more than that.

In an unpublished fragment which apparently was intended as an additional critical reply to one of the essays in the book, *Albert Einstein, Philosopher-Scientist* (1949), Einstein returned once more — and quite scathingly — to deal with the opposition to this view. And now, his word indicate explicitly and with clarity that the change that had begun half a century earlier in his epistemology was now complete. Perhaps even without consciously remembering the words of Planck's attack on Mach cited earlier — that a basic aim of science is "the complete liberation of the physical world picture from the individuality of creative intellects" — Einstein refers to a "basic axiom" in his own thinking: "the postulation of a 'real world,' which so-to-speak liberates the 'world' of the thinking and experiencing subject. The extreme positivists think that they can do without it; this seems to me to be an illusion, if they are not willing to renounce thought itself."

Einstein's final epistemological message was that the world of mere experience must be subjugated and transformed by fundamental thought so general that it may be cosmological in character. To be sure, it is now accepted by physicists the world over that one must steer a middle course between the Machist attachment to empirical data as the sole source of theory on one hand, and the aesthetic-mathematical attachment to persuasive simplicity and internal harmony as the warrant of truth on the other hand. But by going in his own philosophical development from one end of this range of positions to the other, Einstein has helped us to find our own.

Acknowledgement

I wish to acknowledge the financial support for cataloging the documents in the Archives at Princeton, kindly furnished by the Rockefeller Foundation. The Institute for Advanced Study at Princeton and its Director, Dr Oppenheimer, have been most hospitable throughout this work, including the period which I spent as a Member at the Institute during a sabbatical leave. I also am grateful to Mr Vero Besso for permission to quote from the letters of his father, Michele Besso.

[42] Einstein's change of mind was of course not acceptable to a considerable circle of previously sympathetic scientists and philosophers. (See for example P. W. Bridgman, *op. cit.*)

Drafts of portions of this paper have been presented as invited papers at the *Tagung* of Eranos in Ascona (August 1965), at the International Congress for the History of Science in Warsaw (August 1965), and at the meeting, *Science and Synthesis* at UNESCO House in Paris (December 1965). The foregoing text represents the revised and reviewed version of the latter communication.

INTEGRATION AND DIFFERENTIATION IN THE MODERN SCIENCES

General Evolution of Scientific Knowledge

B. M. Kedrov In the general evolution of scientific knowledge we can distinguish three essential steps: first, nature is seen by man as a total whole with undifferentiated parts. Here all is constant movement and interaction. Science assumes a diffuse character. The whole body of knowledge constitutes a unified science which has not yet been broken down and which is completely dominated by philosophy. This is the period of natural philosophy, before the individual science have been differentiated or, at least, exist only in embryonic form as, for example, during the Alexandrine period. And since the sciences have not yet been differentiated, there is obviously no question of integrating them.

Step number two is characterized by rapid progress in the differentiation of the sciences and by the process of decomposition of total science. This was an essential stage in the evolution of scientific knowledge because, without detailed research, it would not be possible to work out a clear scheme of nature. This process of decomposition goes right back to the early days of the Renaissance when, first the mathematical sciences, then the chemical, physical and biological sciences, and finally the anthropological sciences had their origins.

Here the trend towards differentiation is dominant, and the reason for this is clear: before we can study anything, we have to break it down into its constituent parts. Having once solved this problem, we can then find out how these parts fit together to form a coherent whole. This is equally true of the development of all scientific disciplines. Before the physiology of a living organism can be studied, its anatomy has to be known; hence, anatomy is a prerequisite of physiology. Similarly, chemical analysis must precede chemical synthesis.

But the trend towards integration has not vanished. The leading scholars of the Seventeenth and Eighteenth centuries, as also those of the first half of the Nineteenth, attempted to unite all the fields of science and to work out a system of synthesizing all knowledge. But as the trend towards differentiation was in the ascendant, their efforts merely brought about an addition and not a synthesis of knowledge. The latter did not begin to appear until about the middle of the Nineteenth century as a result of such great scientific discoveries as the law of the conservation and transformation of energy, Darwinism, the periodic table of the elements etc.

From this time on the trend towards integration began to gain the upper hand over the differentiating trend. Nevertheless, this did not occur at the expense of differentiation; on the contrary, it developed on the basis of reinforcement of differentiation. Now at last the differentiation of the sciences is seen to be an important factor in their unification, making use of their interactions and correlations. At the present time this feature of the integration of two opposing trends is becoming more and more clearly defined.

Special features of Twentieth century integration and differentiation

During the Nineteenth century the trend to integration became predominant; yet, the trend to differentiation has not disappeared. True, it does not operate in isolation; it is governed by the former and modifies its domination. Why should this be so? Because, while still favoring differentiation, the new disciplines are today contributing to the creation of a general system of knowledge. In order to persuade you of this, I am going to consider them in relation to the whole body of knowledge and to try to determine the position they occupy.

The fringe disciplines are the best cement of integration. Thus, it was not until the end of the last century that chemistry, physics, biology and geology were clearly delimited. Since then chemistry has acquired a whole train of disciplines whose exact degree of relationship is hard to determine: physical chemistry and chemical physics, biochemistry and organic chemistry, colloid chemistry and macromolecular chemistry, first cousin to molecular biology, geochemistry and biogeochemistry, astrochemistry and crystal chemistry.

The branches, while at the same time providing a method of research, advance the process of synthesis. The mathematical method, universally applied in this century, makes it possible to study the quantitative aspect of things and of phenomena, and even the general structure of the links between them. It is impossible to overestimate the importance of statistical research methods which, by summating large numbers of aleatory events, reveal the internal compulsion behind the contingency.

Among these disciplines, I would single out cybernetics which studies the self-regulating mechanism of living systems and of human activity. Some biologists of a dogmatic turn of mind refuse to allow that cybernetics possesses this faculty of analysing life. "To try to do so," they say, "is to reject the very principles of biology." This is where they are wrong.

Finally, the whole structure of modern science rests upon a foundation of general scientific laws which draw all the natural sciences together and link them with philosophy and the social sciences. So you see that now, in our time, the differentiation of knowledge leads not to the isolation of the sciences but to their theoretical synthesis.

Model of the two trends

I propose to analyse the interaction of these trends by means of a model. It does not claim to offer an exhaustive characterization of the advancement of knowledge; it merely attempts to reproduce its common parameters schematically. Our much-simplified model thus ignores the intrinsic unity of the various aspects of the object analysed; if it does show it, it is in the form of a combination of the spatial properties, external properties, therefore, of a geometrical figure. Nevertheless, limited as it is — and, for that matter, as all models must be — our model can show how the two trends interact by the combination and overlapping of the two opposing exercises of analysis and synthesis.

From the point of view of simple analysis, the aspect of the object-model changes according to the standpoint from which it is viewed. But from the synthetic point of view, the different aspects of the model can be seen to belong to the same object by relationships which can be determined.

It has to be shown that under certain sets of conditions the object-model will be projected in different forms and that its seemingly incompatible projections can readily be reconciled.

Let us choose as our model a symmetrically truncated cylinder. According to the angle it makes with the plane, its projection on the plane may be a circle, a triangle, a square, or all three at once — like shadows projected upon the ceiling and two different walls.

In this case the differentiation supported by a unilateral analysis is summed up by noting the existence of projections which apparently have nothing whatever in common. They exist together but their unity is not seen; and so they are isolated. The integration which depends upon synthesis — and synthesis also takes account of the analytical data — will lead us to the geometrical figure which, according to the angle from which the observer views it, has three different and, in a sense, opposing projections. If the angle at which these projections coincide is determined, and if a single geometrical figure is constructed from these projections, we have a model of the integration and synthesis of the sciences. But before we can achieve such a generalization, we first have to study each projection individually; in other words, the geometrical image of the body will be broken down into its constituent elements, then reconstructed on a theoretical level.

Thus, the movement of science from the one to the many, and from the simple to the compound, may be analysed by means of a model.

Methodological basis of the unity of the two trends

Modern science develops dialectically or, which amounts to the same thing, in a manner which is profoundly contradictory. For the object itself is the seat of internal dialectical contradictions. Lenin said that the object (phenomenon etc.) is "the sum and the unity of opposites". This correlation of the opposing processes of knowledge is the true methodological basis of the trends we have been discussing. We know that integration is effected thanks to the appearance of the fringe sciences and the reflection of dialectical transitions between objects, for example, between various forms of matter or of movement.

Knowledge is "an endless process of revealing relationships and new aspects"; our model demonstrates this, although it possesses only three parameters. Let us suppose, however, that these are revealed at ever deeper levels of the structure of matter. Then our model will express the infinity of the possible decompositions of nature.

The laws of matter provide an example of this. They are deployed within the framework of three categories or three aspects of matter which have been studied consecutively by scientists: properties, composition, constitution. The properties of matter and their external manifestations are the first to become known. Composition is revealed by analysis. Finally, the regrouping of the data previously obtained provides us with the means to work out the structure of the substance, that is, the bonding and the arrangment of its particles.

These three categories are thus the three parameters of the substance and the three levels of knowledge they stand for.

Until 1869 these parameters described the molecular level (study of the chemical substance). The discovery of the periodic law enabled scientists to probe the atomic level. In the twentieth century we continue to explore the deeper levels of the atomic nucleus and of elementary particles. Although for the time being the structure of the nucleus and of the elementary particles remains obscure, our knowledge always proceeds along the same path: properties, composition, constitution.

This is equally true of macromolecular combinations, especially high polymers, and the fields studied by organic chemistry, biochemistry and molecular biology. Here, chemistry and biology flow together — the unity and the correlation of the trends towards differentiation and integration are made manifest in full strength.

The complicated and very special situation we find in the natural sciences arises from the fact that the properties and manifestations of living matter — and heredity, in particular — are shown to be in intimate connection with their material physico-chemical constituents. We have here the same chain: properties, composition, constitution, but with this difference: the biological properties (heredity) are found to have the chemical composition and constitution of complex substances, such as nucleic acids which, as we know, play a vital part in heredity, biosynthesis and many other biological processes.

At present, the trend towards unification of the sciences is being continually reinforced and shows in the highest relief the dialectics of the evolution of modern sciences.

PART TWO

Science and Synthesis: Debates

From Plurality to Unity

François le Lionnais The great discovers can readily be classed under two types of mentality: those who dig deep and those who range wide. Those who possess the gift of combining depth with breadth are rare indeed. Albert Einstein was one of them. I need only remind you of his main contributions to the advance of physics for your minds to be staggered by such creative versatility. First, there are the theories of relativity, special and general, and the unified field theory. In atomistics, there is the theory of Brownian motion and of fluctuations. In quantum theory, there are essentially the theory of the light quanta, or photons, and their contribution to the photoelectric effect, but there are in addition the theory of the specific heat of solids, important laws of photochemistry, stimulated emission as applied in the laser, Bose-Einstein statistics... You see that this list contains an impressive number of subjects, at least half of which might have qualified for various Nobel prizes; and, even so, I have not mentioned his minor inventions and discoveries, such as the inverse gyromagnetic effect and many more.

For the general public Einstein is primarily the father of relativity; however, this was not what won him the Nobel prize in 1921 — it was his theory of photons or light quanta. This theory, which postulated the existence of photons, considered them to be both particles of light and electromagnetic waves, and thus opened up a new road along which we have since travelled a very long way. It was Mr. de Broglie's privilege to complement this revolution in the field of light by a no less devastating revolution in the field of matter and the links which exist between these two fields: this is wave mechanics on which and from which we still live today. I can think of no one more appropriate than the father of wave mechanics himself to recall for us the virtues in this field of the father of photons, to retrace the path he took and to show us where these two theories are linked. There is no doubt that both will remain permanent monuments to scientific thought.

ALBERT EINSTEIN AND THE CO-EXISTENCE OF WAVES AND PARTICLES

Louis de Broglie In speaking of Einstein, it is usual to mention only that he was the inspired author of the theory of relativity, first in its special and later in its general form. To apply such a limitation is to overlook the fact that, in so doing, we perpetrate an injustice to a large and very important part of his work. Indeed, although the theory of relativity was and is one of the outstanding triumphs of twentieth century physics, it should never be forgotten that Einstein, with his theory of the "quanta of light," was also the first to suspect the co-existence of waves and particles in light; and again, by introducing new and sometimes bold methods into statistical thermodynamics, he was the first to deduce from it the general theory of fluctuations and of Brownian motion. Of the three basic contributions he made to theoretical physics, the second (theory of light quanta) is just as important in our eyes as the first (theory of relativity), whereas the third put forward original ideas which have proved extremely fertile. And not the least amazing feature of this success story is that Einstein made all three of these magnificent discoveries public within a single year (1905) when he was only 26 years old.

There is, moreover, a clear relationship between these three previously unknown fields which the great physicist explored simultaneously. The fundamental ideas of relativity, and in particular the principle of the inertia of energy and relativistic dynamics, constantly gave him his bearings in his study of the particle constitution of light, and it was his thorough familiarity with statistical thermodynamics which provided him with many of the proofs substantiating the existence of light quanta, which we now call photons, particularly in his work on energy fluctuations in blackbody radiation, on energy exchange and quantity of movement between atoms and radiation in a field in thermodynamic equilibrium, as well as in his analysis of spontaneous and stimulated emissions which — more than 30 years ahead of its time — provided the key to the understanding of those useful pieces of equipment we call masers and lasers. We see, therefore, that there is an underlying unity in the research Einstein carried out during the early part of his career.

I propose to look first at that part of his life's work which falls within the period from 1905 until the outbreak of war in 1914. It must have been about 1903–4 that the young patent examiner in Berne began to think, with that concentration of thought often conferred by working alone, about the problem of quanta; he was at once struck by the paradoxical character of the hypotheses which, not long before, had enabled Max Planck to deduce the exact formula for the spectral distribution of energy in blackbody radiation. He was convinced that radiation emission and absorption must be inverse processes, and this he thought impossible to conceive under the classical view of light as pure waves; he was thus led to suppose — going further than Planck had dared to go — that the luminous energy emitted by particles must retain its particulate form during wave propagation. Recalling then that the photoelectric effect, discovered almost 20 years before by Hertz and Hallwachs, constituted an insoluble enigma for Fresnel and Maxwell's wave theory of light, he postulated that luminous energy of frequency v is transported by particles of energy hv, h being Planck's constant; he then called these particles "light quanta", though we now call them photons. From this point, by a two line calculation, he derived the explanation of the mysterious phenomenon which had until then disconcerted all the theorists.

Let me say right away that the explanation of the photoelectric effect supplied by Einstein at this time, after being for a long time subject to doubt, was in the end triumphantly vindicated by experience, first by Millikan for light in 1916 and then, a short time later, by my brother and Ellis for X-rays and gamma-rays. Compton's discovery of the effect which bears his name, and the theoretical interpretation of this effect by the photon hypothesis, soon after given independently by Compton and Debye, completed the confirmation of the correctness of Einstein's ideas. And it was specifically for the discovery of the law of the photoelectric effect that he received the Nobel prize for physics in 1921.

But although Einstein's idea was to triumph in the end, at the time he put it forward a chorus of protest arose against it, as is often the case when a fundamentally new way of looking at things violently overthrows the old habits of thought. It was objected that classical optics had had its successes: the phenomena of interference, diffraction and polarization had since Fresnel, and with the improvements subsequently introduced by Maxwell's immortal work, really appeared to confirm the identification of light and all electromagnetic radiations with wave phenomena. Now, Einstein did not quarrel with this; on the contrary, he dreamed of reconciling these two points of view. He said to himself: "Why should one not assume the existence of light particles, sources of radiant energy, which would be transported by an electromagnetic wave of zero or negligible energy, yet in such a way as to lend to luminous energy the statistical distribution envisaged by the wave theory?" This was the "ghost wave" or *Gespensterwelle* theory, to which I have devoted much thought in recent years. Einstein also noted that this would not prevent luminous energy, converted into protons by an atom or a molecule, from being projected in a definite direction, and it was this which led him to the concept of "needle radiation" or *Nadelstrahlung*.

For several years — from about 1905 to 1910 — Einstein was preoccupied with seeking proofs for the coexistence of particles and waves in light. Making use of his thorough knowledge of statistical thermodynamics, he established the formula for energy fluctuations in blackbody radiation and showed that it can be broken down into two terms, one of which corresponds to the existence of particles and the other to the existence of waves in light. I have in the past given a great deal of thought to this strange formula and I have recently taken up the study of it together with my young coworker. Mr. Andrade e Silva, because I consider it capable of giving us valuable information on the true nature of the coexistence of waves and particles, at least, in the case of the particles we call bosons and of which photons are a part.

In the same order of ideas, Einstein, first alone and then with Hopf, studied with the aid of clever thermodynamic analyses the Brownian motion of a particle which, when placed in a blackbody radiation, constantly exchanges energy with it. Again he found a two-term equation corresponding to the existence of waves and particles in blackbody radiation and showed that Planck's law alone could explain the predicted fluctuations.

Let us now pass to the research Einstein carried out on similar ideas during the period 1917–1927. During this time, Einstein was at first deflected from his work on light by the efforts he had to make to give his theory of relativity a general form, and this led him, as we know, to a brilliant and memorable interpretation of the field of gravitation. However, he did not lose sight of the quanta and continued to think deeply about the foundation of the atomic theory which Niels Bohr had developed in 1913 and about the more sophisticated form given to it later by Sommerfeld and to which Einstein himself made a significant contribution. This brought him in 1917 to a brilliant return to his earlier studies in the famous memoir where, introducing instead of atoms the idea of spontaneous and stimulated emissions, he showed that there is a hidden relationship between Bohr's law of frequencies and Planck's law of blackbody radiation. I have already stressed the importance assumed in recent years in physics by the idea of stimulated emission. To these well known conclusions Einstein added in the same paper some new considerations on the Brownian motion of a particle immersed in blackbody radiation and demonstrated that the photon, having energy h, should possess a quantity of movement. This was the foundation of the dynamics, necessarily relativistic, of the photon; in 1922 I was to study various aspects of this dynamics in an article in the *Journal de Physique* and two years later in my doctoral thesis.

I hope I may be forgiven for bringing in my own work at this stage of my talk. It was, as a matter of fact, at the end of the 1914–1918 war that, with a mind impregnated with Einstein's ideas, I began to think seriously about the problem of waves and particles. I had already done some work on this previously and I remember when I was preparing for an examination in rational mechanics at the Sorbonne being greatly struck by the Hamilton-Jacobi theory and by the analogy between Fermat's principle and the principle of least action. But it was not until 1922 that, as a first draft, I published an article on blackbody radiation conceived as a photon gas, and a note in the *Comptes Rendus* on the Einsteinian formula for energy fluctuations in blackbody radiation. Then suddenly, in the summer of 1923, I had the idea that the coexistence of waves and particles which Einstein had discovered in light might be equally true for all particles of matter. I thereupon wrote some notes for the *Comptes Rendus* of September 1923 and later, in my doctoral thesis of 1924, expressed in a more complete form what were to become the basic principles of wave mechanics.

When I began my research, Einstein did not know about it, but he had at that time just heard about the work of the famous Indian physicist, Bose, who developed a new form of statistics which could be applied to photons — which, incidentally, I had also glimpsed in my 1922 paper. Einstein was extremely interested in Bose's work and undertook to develop it in a paper presented to the Berlin Academy of Science. It was then he heard from Paul Langevin about the manuscript of my doctoral thesis. He recognized at first glance the naturalness of generalizing his ideas about the coexistence of waves and particles. He gave Langevin a very favorable opinion on my thesis and mentioned it in the last of of his notes on the statistics which have since that time been known as the Bose-Einstein statistics. Erwin Schrödinger heard about my theory through Einstein's notes and thus in 1926 was inspired to write his celebrated papers on wave mechanics. Not long after this, in 1927, the discovery of the phenomenon of the diffraction of electrons fully confirmed the ideas I had put forward in my thesis.

Now we come to the time when Albert Einstein, more and more preoccupied with working out his ideas of a unified concept of the electromagnetic and gravitational fields, hampered in his work by the trials and tribulations which arise from celebrity and very probably, too, worried about the turn of events foreshadowing the Hitler regime in Germany, where he had long been settled, ceased to make worthwhile contributions to the problem of the coexistence of waves and particles. But it was about this time, too, that in seeking to understand how matter enters into gravitational fields, he began to look for singular solutions, or solutions at high local field concentrations, to the equations of general relativity representing particles of matter.

I was certainly more or less consciously under the influence of Einstein's work when in 1926–1927 I developed my interpretation of wave mechanics which I called the "theory of the dual solution", published in the *Journal de Physique* for June 1927. I had always had the idea, and I believe it to be in conformity with Einstein's, that the dual particle/wave aspect of light and matter implies the real and simultaneous existence of very intimately linked waves and partieles. Hence I wanted to construct a clear and precise picture, and this was what brought me to my still very imperfect conception of the theory of the dual solution. In an academic note entitled "The dualism of waves and particles and the work of Albert Einstein", reproduced in my book, *New Perspectives in Microphysics*, I explained in detail how I put my theory — unfortunately in a greatly abridged form — to the Solvay Council of Physics which met in Brussels in October 1927, and how the physicists of the Copenhagen School, which had sprung up around Niels Bohr, countered my thesis with an interpretation based on the ideas of complementarity and uncertainty. Einstein was the only person who encouraged me; [1] although he made justiable reservations about the way I had presented my ideas, he told me, more or less, "Do not worry, you are on the right track."

After 1927 the probabilistic interpretation of wave mechanics formulated by Bohr and his school rapidly gained ground, and this led to an increasingly abstract presentation of "quantum mechanics", and then to the "quantum field theory". In this almost nothing survived of the physical ideas about waves and particles which had guided Einstein in his theory of light quanta and had later inspired me to make a bold generalization of these ideas, and which constituted wave mechanics. To be exact, there are no longer either waves or particles in the interpretation of the Copenhagen School. In effect, they said particles were probably present in a whole region of space, which could be very large, and this is the utter negation of any clear conception of a particle. As for the wave, it was arbitrarily standardized so as to become nothing but a representation of probability, and thus we are no longer able to understand how the wave can determine such physically observable phenomena as interference, diffraction or steady-state energies of the atom, because a representation of probability could not conceivably be the cause of a physical event. A mortality table can show most exactly how many persons are likely to die in Paris next year, but those Parisians who do die next year will die because of illness or accidents, or some similar cause, and certainly not in obedience to the mortality table.

Neither Einstein nor Schrödinger ever accepted this interpretation of quantum physics which was pushed at them. But Schrödinger wished to retain just the classical wave concept and to discard the idea of particles, and I never thought he was right in this. Einstein, on the other hand, wanted to retain the physical view of both waves and particles. He declared that the formal concepts of the "orthodox" theory, while no doubt giving precise statistical representations, did not present a complete picture of physical reality. His objections to the theory as it was current at that time and the inconsistencies he found there during his polemics with Niels Bohr will be found in his "Reply to Criticisms", which is printed at the end of the jubilee volume dedicated to him on his seventieth birthday, and also in the very remarkable "Remarques préliminaires sur les concepts fondamentaux" which he did me the honor of writing as an introduction to the book compiled to celebrate my sixtieth birthday. In the latter he particularly stressed the fact that the theory he dreamed of

[1] During the Solvay Council discussion, Einstein raised a most penetrating objection to the *exclusive* use of waves in wave mechanics and concluded by saying, "In my opinion this objection can be met in only one way, by describing the process, not as a wave alone, but by localizing the particle within the wave during its propagation. I think Mr de Broglie is right to look in this direction. If we speak only of waves, the interpretation of ψ^2 to my mind implies a contradiction of the postulate of relativity."

would probably be based on non-linear equations and ought, of course, to introduce a kind of Brownian motion of the particles.

You will readily understand that when Einstein heard in 1951–1952 that, after having put them aside for a long time, I had again taken up the ideas of my youth and that I was tackling the difficult task of resuming my attempt at interpreting wave mechanics by the theory of the dual solution, looking into it with the aim of perfecting it, he was delighted and sent me his best wishes. He also sent me by the hand of a friend one of the last photographs of himself ever taken, as a mark of appreciation.

In 1960, five years after Einstein's death, I had become aware of the fact that the theory of the dual solution could only be perfected by introducing a random element to explain the constant intervention of probabilities in wave mechanics, and that a particle, however isolated it might seem to be, must be in constant contact with a hidden medium constituting a sort of thermostat. This led me to accept the existence of the "quantic medium", postulated by Bohr and Vigier as long ago as 1954. Taking up again in a new form certain ideas I had had some 10 years earlier, I then developed a "hidden thermodynamics of particles" which I explained fully in a book published in 1964 by Gauthier-Villars and summarized in an article which appeared in the *Annales de l'Institut Henri-Poincaré*. This new theory has three basic principles: first, of course, the coexistence of waves and particles, represented by the theory of the dual solution, for example, in the case of light, by accepting that photons are conveyed by a wave of negligible energy, analogous to Einstein's old "ghost-wave"; next, the general ideas of the theory of general relativity and in particular the principle of the inertia of energy and the form assumed by thermodynamics in relativistic conceptions; and finally, the introduction of a kind of Brownian motion of particles and the study of the resulting fluctuations by procedures inaugurated long before by Einstein. Thus, this undertaking, one of whose most remarkable conclusions is to derive the principle of least action from the second law of thermodynamics, draws together into a single whole the three great principles which exactly 60 years ago sprang almost simultaneously from the highly original brain of the greatest theoretical physicist of the century.

And since, in my recent work as well as in my young days, I have always been more or less consciously guided by the powerful thinking of Albert Einstein, it was perfectly natural that I should wish to add my quota to the homage rendered to him, and that I should seek to bring out the importance of some parts of his work which have been allowed to fall into oblivion.

Bibliography

On Einstein's work on the coexistence of waves and particles, see the exposition by Martin J. Klein "Einstein and the Wave-Particle Duality" in: *Natural Philosophers*, New York, Blaisdell Publishing Co., 1964.

On Einstein's ideas about the interpretation of wave mechanics, see his article "Remarques préliminaires sur les concepts fondamentaux" at the beginning of the book *Louis de Broglie, physicien et penseur,* Paris, Albin Michel 1953.

On the theory of the dual solution, and the hidden thermodynamics of particles, refer to the following books: *Etude critique des bases de l'interpretation actuelle de la Mécanique Ondulatoire,* Paris, Gauthier-Villars, 1963, and *La Thermodynamique de la particule isolée,* Paris, Gauthier-Villars, 1964, where a more complete bibliography will be found.

Reverend François Russo I have no intention of giving you anything like a systematic account of it, but I would like to stress the importance of the contents of the famous memoir of 1905 which introduced special relativity. Unfortunately, very few people, even educated people, are capable of understanding properly why this date is so epoch-making.

In order to understand it, we have to look at it from two points of view: the objective point of view of the effective result achieved, and the subjective point of view of the merit of the man to whom we owe this result. The objective point of view is familiar and I shall not dwell upon it; but I would like to go more deeply into the subjective one. What was Einstein's greatness in making this discovery? First, we must be careful not to attribute to him things that had already been discovered. This is something we should always bear in mind when speaking of great men, because there is a tendency to credit them with the discoveries of their predecessors. Einstein said himself in his autobiography that in 1905 the time was ripe for the discovery of special relativity. It was ripe in a very precise and perfect sense. The fruit was just ready to fall. Why was this so? Because physicists were confronted with a singularly difficult and baffling problem: the impossibility of reconciling mechanics and optics. I shall not remind you of everything that had been done since the failure of Michelson's experiment; all the difficulties had been identified and the problem had been correctly defined. The attempts to solve it had been numerous and a mathematical structure — the Lorentz equation — that seemed tailor-made for the theory of relativity had been worked out. So we may say that by 1905 the field had been swept and garnished.

So where does this leave Einstein and his exceptional merit? Ah, but however much all was ready and waiting, there was still one more step to be taken and Einstein was the only man able to take this step. And what was so singular about this step? It was primarily methodological. It consisted first in re-examining the generally accepted fundamental ideas, that is, the ideas then current, about time, simultaneity and the ether. This re-examination arose from the fact that Einstein could never tolerate being unable to explain with perfect clarity any question in which he was interested.

But the essential point of this procedure is the positive, not to say positivist, attitude it betrays. It is this point and this attitude that I wish to emphasize. Now, of course, positivism was fashionable and widespread well before 1905. Einstein had let himself be carried along by its flow; he himself said what a lot he owed to a philosopher, Hume, and to a scientist, Mach. But Einstein's particular merit was to have taken positivism to its logical conclusion. Before 1905 positivsm had had an admixture of very un-positive ideas; this is not said often enough. But Einstein's positivsm was radical and total. Moreover, he did not, like the non-scientific philosophers, just talk about it, he put it into practice, rejecting ideas which had until then been accepted and insisting that any fundamental idea must be capable of being defined in a positive manner, that is, on the basis of observations.

Later Einstein was to claim that he had discarded this positivism; you will no doubt remember how he replied to those who said to him at the time of his notorious debates with Bohr, "But how is it that you, a leading positivist, who have been so successfull thanks to this positivism, now reject the positivism of those who defend wave mechanics and indeterminism?" You know how Einstein replied, "A good joke ought not to be repeated too often" — I don't think he was really quite fair to himself with this reply, as at bottom he always remained a positivist. But later on he tried to step back and take a long view of positivism — we shall be hearing more about this presently; but even in 1905 when his memoir on relativity came out, Einstein was not the limited kind of positivist, tied to the facts and content just to co-ordinate them. His way of proceeding stems from a refusal to be impressed by realities which he has not checked up on himself, and this is positivism transmuted and shot through with rationalism, and infinitely more constructive. With Einstein, special relativity consisted of deliberately and by a very lucid act of intelligence laying down two principles: that of the validity of the laws of physics in all the reference trihedra in rectilinear and uniform displacement; and that of the constant speed of light. Nevertheless, I think one may say that at the heart of Einstein's procedure there is an attitude of a methodological kind. Einstein's vigor — and his merit — derived from assuming an attitude that was both positive and radically positivist. I think it was this attitude rather than his scientific ideas which constituted his strength.

One may also agree with Father Dubarle that Einstein's attitude is basically classical in the sureness and firmness of its intellectual processes, the sureness of a mind which is not swayed by emotion or impressed by accepted ideas and which recognizes only the cold processes of logic, checked and cross-checked. This is why the discovery of special relativity alone, quite apart from his subsequent astonishing and perhaps even more original discoveries, strikes us as an epoch-making contribution not only towards the understanding of physical phenomena but also towards scientific method — a major step in moulding the mental attitude by means of which we may hope to penetrate to the reality of things.

Jean Ullmo I do not altogether agree with my friend, Father Russo, in his opening assertion that in 1905 the time was ripe for the fundamental discovery of relativity. But I do agree that Einstein started out from positivism, that is, with the rigor of never being satisfied with concepts, however self-evident they might appear to be intuitively. I think this attitude is fundamental to science: one should never accept anything to begin with. No concept has an absolute value in itself, it should be investigated afresh every time; and the only way of re-investigating it thoroughly is to find out whether it works, in other words to see whether it is possible to define it in an exact manner by a regular and repeatable series of experimental operations.

However, I think that what was fundamental with Einstein was not this necessary precondition but, on the contrary, his complete reversal of perspective. He upset the rather lazy idea — or, at least, the customary one — of science as a collection of recipes expressed in the form of laws which confer a certain power of action and prediction. This is an attitude which flowered in its finest form in the conventionalism of Henri Poincaré.

Indeed, in this context, the example of special relativity is particularly instructive because in 1905 Henri Poincaré was for his contemporaries the grand old man of both physics and mathematics. No-one today will deny that he dominated his age with his intellectual power, both as scientist and thinker. Now, Poincaré had trained the whole of this power on to the difficulties of the time, the interpretation of Lorentz' equations and the baffling phenomena of the contraction of length and the expansion of time. Poincaré was totally incapable of effecting any sort of mutation in interpretation because he was inspired by a false philosophy — the recipe variety, with its conventionality, its procrustean frame within which all phenomena had to be accommodated, with the use of a little force if need be. I believe this to be the earliest example of an obstacle placed in the path of science, the delaying of an urgently needed scientific discovery by a wrong philosophical attitude; however, I think there have been a number of examples since then, and nothing is more misguided than to suppose that in the pursuit of science one can forget about philosophy, epistemology and methodology. It is this sort of thinking that puts a man a whole generation behind the times. We have seen some prime examples of this — indeed, they still occur.

What, then, was so new about Einstein's attitude? It was his belief in the reality of the physical theory he was working out. Let us start with the example of special relativity. He believed in the reality of his new space structure — space-time structure, if you prefer. He set this reality against the old model which, since Kant and even before, had appeared ontologically evident, the model of absolute space, Newtonian and Kantian space in which mankind had been bathing intellectually since the seventeenth century, as was so admirably demonstrated by my late-lamented friend, Alexandre Koyré. Einstein rejected out of hand the idea of mankind's presence in absolute space and equally absolute intuitive time. He told us that the space and time surrounding us do not have the structure we supposed, that very structure which Kant considered so obvious he made it into one of his categories of thought. Was this not revolution indeed? Can one really assert that the time was ripe for a revolution as radical as this? It is probably the greatest mutation ever in the history of thought.

Let us now try to understand what it was about these ideas of relativity which constituted Einstein's glory but which were essentially misunderstood. Einstein was the man of relativity, but to the man in the street relativity denotes scepticism, indifference, a negative and defeatist attitude. But with Einstein, on the contrary, relativity is the very prototype of the positive — and I mean positive and not positivist — attitude, the affirmation of a reality of which he was certain. From start to finish Einstein's relativity aims at distinguishing the objective, the non-relative fact. Relativity is a means to attain the end, which is objective reality.

Space is indifferent to the phenomena which occur in it, space is neutral, a container in which physical objects are made manifest by their interactions but which does not act upon them. However, space has a certain structure and it is not the structure which had previously been postulated: Newton's Euclidian structure, which I mentioned a while back. Once the old structure is done away with, a whole series of physical manifestations are now seen to be interactions between objects and are thereby explained, dissolved one might almost say, by the very fact that they are revealed to us by the structure of space, though, of course, space has nothing at all to do with it.

Take, therefore, as a typical example, the Lorentz contraction. Much ingenuity had been expended over the years in looking for physical phenomena due to speed which could explain why distances should be shorter in a moving system. At a single stroke all this seeking for non-existent physical properties is disposed of — suddenly, all is clear, it is merely a matter of the intrinsic properties of the dimension, and hence of the definition, of space and time.

This, then, is the first principle: special relativity is the affirmation — and the consequences which flow from this affirmation — that what happens in space depends upon the structure of space in the sense that this defines the measurements we carry out. But, of course, our measurements do not change things in the least, and the phenomena which go on in space are not affected by them; this is the principle of relativity.

Oddly enough, the second principle of relativity, that of general relativity, appears to be the exact opposite of the first. Here, on the contrary, space is no longer an indifferent container, it is itself an object reacting with other objects in the world and has a physical structure of its own. This structure explains the basic phenomenon of gravitation and ever since people have tried to use it in attempts to construct unified theories which will explain the totality of other phenomena which make up the material world.

We may ask — and this is my last point and I shall try to sketch it very rapidly — how it was possible to use the same name, principle of relativity, to describe two opposite attitudes, one the neutrality of a container space not acting upon the objects it contains, the other the virtually absolute position of space which has a structured form and which does act upon objects.

In general relativity the planets describe certain trajectories in exactly the same way as a marble thrown upon a surface will assume the shape of it and describe a geodesic curve. The shape of the trajectory of the planets corresponds to the shape of the geodesic curve, that is it is the shortest distance between two points on a curved surface. This is how space intervenes and acts in general relativity.

How does it happen — for I think this has been and still is the source of very considerable misunderstandings — that both these statements can be designated by the same word: relativity? The factor they have in common is the idea of the transformation of the system of reference. I defined special relativity a little while ago. It is a corollary of the neutrality of space relative to the events and phenomena going on in it that these phenomena must be independent of the frame of reference within which one positions onself in order to measure distance and time. So here we have invariance relative to changes in the system of reference. But it has to be understood that, where there are two different ideas, the second can be a consequence of the first, but the first is not necessarily a consequence of the second; this is an implication in the logical sense. If space is neutral, the phenomena must be independent of the system of reference, but the reciprocal is not true.

Now, it is this implication which is retained in general relativity. No doubt Father Russo would say that this is a methodological condition; for myself, I see much more in it — a fundamental philosophical condition, a supremely rational condition, in fact. Let us remember that the phenomena should be independent of the system of reference applied to them, because basically what is being questioned is the very idea of objectivity and not at all the physical idea that space is neutral and has no effect. It is quite a different matter to say that the system of reference does not act upon what is being measured and to say that space as a container does not act upon the objects it contains. If we retain the first statement and reject the second, we have general relativity. General relativity is the simple affirmation that, whatever the laws of physics may be, they should not depend upon the way we determine them, the way we fit them into a formal framework. I do not think it has been a complete waste of time to underline these two points.

André Lichnerowicz I would like to add a few brief comments to define exactly what these differences are. I shall certainly not repeat what has been put so well. I think Einstein's destiny is peculiar in the sense that it was perhaps in 1915 with general relativity that he became the latter-day Newton for all men all over the world. If he had written only his 1905 paper, he would have been one of several theoretical physicists of genius, but he would not have towered above such men as Lorentz or Minkowski, except by his scientific courage, which was perhaps his most outstanding quality. But in 1915 with the publication of his theory of general relativity, which I prefer to call the relativistic theory of gravitation, he put himself right out in front.

There is something rather strange about this theory, because it is primarily a theory of what I would call the mathematical representation of gravitation. A few experimental phenomena are predicted by the theory; they are very limited, a mere three or four (the perihelion of Mercury, the red shift, the bending of light rays). Clearly, it was not these which aroused such tremendous interest in relativity, but the intellectual content of the theory. For the first time since Newton the intellectual content had pride of place, not merely for specialists, but for everybody. The Einsteinian theory of gravitation is thus of extreme importance for the scientific future of mankind. What is realized here is the finest example of a physical theory as constituted in the very conception proposed by Einstein himself.

Instead of talking about positivism, I should prefer to say that a physical theory always consists of a mathematical substrate, which is the "meat", and a superstructure which is just padding. The padding is always concerned with "idols", and Einstein has destroyed more idols than anyone else by insisting on just two things: mathematics and direct experimentation, even though the latter is "all in the mind".

A physical theory, in Einstein's conception, springs from the free creative activity of a man who sets up axioms to start with and need only justify them by their results, which are sometimes rather distant, and by a conviction of internal coherence when the proposed theory unites very wide areas of physics. In the relativistic theory of gravitation it is perhaps the power of understanding rather than the power of predicting new phenomena which has long attracted attention. The current interest in the relativistic theory of gravitation, which has revived over the last ten years, is now much more orientated towards experiments. Following a number of experimental and theoretical discoveries, we perceive that phenomena which used not to be observable are now in process of becoming so, and we wish to use them to verify the theory of gravitation.

I do not quite agree with Mr. Ullmo on one point. I dislike his definition of general relativity as the invariant relative to changing coordinates; it could be said that any physical theory could be put into a form which has this character. It is relatively easy to take the most classical dynamics as described by Lagrange and give a description of it which is independent of the dimensioning of space-time configuration.

General relativity gives the field of gravitation a very special part to play because — under its name of curved space-time — it serves at one and the same time as a background or general frame for all the physical phenomena which go on within it, and yet is simultaneously shaped by these physical phenomena themselves. It is perhaps this duality which lies at the origin of the ambition all physicists have to construct a unified theory. Only nowadays the unified theories are programs, mathematical research jobs, but physics — never!

The unified theories of Einstein's ambition were devised to avoid a scandal. The scandal was that his equations had two parts: the left side had a respectable geometrical ancestry describing the curvature of space-time and the right side, the impulsion-energy tensor which is the source of the gravitational field, had only a phenomenological ancestry — rather suspect, this, from the scientific point of view. This part included the equivalent of an electromagnetic field — when there was one; so here we had this electromagnetic field, correct from the point of view of Maxwell's equations and from many other points of view, arising in the second member to bend time-space into a stature unworthy of it. Obviously one could look for a more general geometrical object than Riemannian geometry (a geometry already prepared by Ricci and Levi-Cività, as it might be specially for Einstein) which simultaneously describes the field of gravitation and the electromagnetic field. And this, which was the subject of much research by Einstein at the end of his life, finally produced a theory, called the Einstein-Schrödinger theory, which is of very great interest, yet it is hard to say what it represents or what it might represent. I should say that the actual notion of the unified field theory evolved simultaneously; there were few physical fields known to be associated with particles and the idea itself underlies Einstein's efforts. But gradually, now that we have meson fields and all the fields of all the spins at our disposal, although the problem of a unified field theory remains unresolved, at least we now know that the difficulties are much greater than one might *a priori* suppose.

I wish to conclude with a point which bears on one of the difficulties encountered by Einstein himself at the time of the subsequent development of general relativity. The tool of local Riemannian geometry was created before Einstein came on the scene; the appearance of Einstein's theory stimulated fresh work on local Riemannian geometry, but what was really needed was a *global* geometry. Even on the non-cosmological scale, when one is working on the scale of the solar system or of a galaxy, the solutions for which one is looking are not local problems, they apply to all the space-time in which one is interested, and so a study of the global geometry of this space-time is needed.

Now, no such geometry existed; it began to take shape around 1926–1927 with the work of Elie Cartan. But for ten long years general relativity and Einstein himself lacked an essential tool which the mathematicians had not yet developed, but which now supplies us with new procedures. General relativity was born of Einstein's postulates and we are now able to express these in a suitable form. It remained — and this was done above all by Einstein and his co-workers — to progress from a state equivalent to Newton (1915 paper) to the equivalent of the work of Laplace or Lagrange. Einstein, Infeld and Hofmann on the one hand and the great Russian theoretician, Fock, on the other, worked out around 1936–1937 the first methods of what one might call relativistic celestial mechanics. Of course, the results obtained are extraordinarily small and astronomists in most cases do not have to consider them; but we are beginning to be able to handle problems of this kind and, indeed, we shall have to learn to do so. Satellites and atomic clocks beckon us on. This shows how very up-to-date Einstein is.

Jean-Pierre Vigier I want to make three brief comments on what has been said. I should like to begin with Mr. de Broglie's paper. I think he is basically right to insist on the apparent duality but underlying unity of Einstein's work. All living thought creates scandal and today the problems to which Einstein devoted his life are more up-to-date than ever they were. On the one hand, he attacked the problem of the continuum in nature, trying to combine all the laws of nature into one unitary geometrical theory; on the other hand he was the origin, the deep source, of the quantum theory. This contradiction between the continuous and the discontinuous is always at the core of the problem physicists have to face today when they consider, for example, the theory of elementary particles.

For a time the tide of physics ran against Einstein. Relativity had made its way in the world, but his ideas about quanta ensured him a splendid isolation at the end of his life. All the younger generation of physicists had turned away from him. Einstein's attempted synthesis, representing matter as single units within fields, remained restricted to the unitary theories and the unification of the fields of gravitation and electromagnetism.

To render a true account of Einstein's greatness, we must look at the development of physics since his death. It has evolved in two directions. However, before I go into this, I should like to add a word on the nature of his procedure. I think that at bottom Einstein was an anti-positivist. Poincaré had all the elements to hand for constructing special relativity. The group which bears Poincaré's name was already written and Maxwell's equations could be interpreted. Einstein's courage consisted precisely in going against Mach, in breaking with him, even if he did not at first recognize this himself. He broke with him on two fundamental issues: first, he was bold enough to say that the fundamental approach of the positivists, who deny the reality of anything which is not directly measurable, was in error. He proclaimed the objectivity of the Lorentz group and of the laws of nature which are independent of the observer, and he constructed a theory which was at the opposite pole from Mach's whole attempt at mechanics. The next phase was mentioned by Father Russo; I think it expresses very clearly how Einstein himself evolved during the course of his life.

I should now like to return to my analysis of the evolution of physics since Einstein, with the aim of showing the extent to which the problems he grappled with have remained current and how at bottom his thought still provokes scandal and strife.

Our conception of matter has been enormously enriched by his work. A vast number of elementary particles have irrupted onto the scene. The giant accelerators are always turning out new ones. If you talk to an experimental physicist, he is sure to tell you of new resonances or discuss their possible existence. On the other hand, we know that one property these particles have is the ability to change into other kinds of particle, and we are all looking for some symmetry, for properties capable of explaining these transformations and recreating unity out of all this disorder.

On these two points Einstein's thought, as explained by Mr. de Broglie, lies at the very center of the modern polemic. The two questions he put are still very much alive. *First question:* Is the description of nature supplied by the quantum theory as it now stands complete, or should we look at deeper levels, the sub-quantic level, for instance, for new realities which may lie hidden beneath the actual phenomena and which could explain them? This is exactly what Einstein himself did in his time to explain, with Smoluchovski and others, Brownian motion. *Second question:* Is there a possible unitary theory behind this disorder, or must we be content with phenomenological descriptions, recipes which enable us to distinguish partial categories among these phenomena?

These two approaches enabled Einstein to make his discoveries, they dominated the argument which raged about him in his last years and they are still very much alive. There is, first, the attempt to look behind the actual disorder of particles at the deeper levels of nature which might permit the restitution of order and causality in the apparent disorder we observe, and, second, the attempt to unify and geometrize the physical fields. I believe nobody so far except Mr. Lichnerowicz (and I think he was right to stress the point) has analysed the importance of Einstein's essential procedure which was his attempt to geometrize the laws of nature. Naturally, since Einstein we have discovered other fields. In his day, only the so-called long-range fields were known — gravitation and electromagnetism. The fundamental contribution of the last few years has been penetration into the domain of high energy and the discovery of short-range fields. This raises the problem (raised, indeed, by Mr. de Broglie) of working out a way of describing the structure of particles within a new geometrical framework.

In this connection, I should like to give you my views on Einstein's essential idea: his attempt to describe the laws of nature, not within the framework of classical space-time, but in terms of movements describable completely within the framework of space and time. This procedure does not mean there is no link between space and time. The problem we have to face at present arises out of the fact that, with the experimental methods available to us, we have explored the particles in a highly superficial manner. We have observed the appearance of new quantic numbers and new particles, but at present we lack the tools — though we shall have them in the future — capable of determining whether or not these particles have a non-punctate structure within the framework of space and time, or whether they are in reality obeying non-linear equations. It is clear that the theoretical approach must rest upon experimental discoveries, but it must also rest upon a methodology. Now, Einstein's methodology, the philosophy which inspired his actions, that is, the idea of the objective character and material nature of phenomena, the possibility of describing properties in terms of space and movement, and above all the possibility of constructing a more fundamental dynamics able to account for the essence of experience, this particular action (which is quite a different scientific battle order from the one adopted by most physicists today) is still the center of debate and controversy.

Is there a simple reality behind the apparent disorder of things? Einstein used to say that God is a mathematician and does not play dice with nature. He had a very deep-seated confidence, only to be proved or disproved by experimentation, in the rationality of things and in man's ability to understand what goes on, to move forward step by step in his understanding and unification of natural phenomena. Behind the disorder, he hoped to discover the invisible simple answer, and to lift the veil which hides the underlying reality of things. His ideas of the geometrization of nature, his new fundamental dynamics, these are still at the heart of research today.

I am ignorant of the outcome which the history of physics will award to the present situation. One thing is clear, however, and this indeed shows the greatness of Einstein, he did not sit on the fence, he came down on one side in the battle, and his ideas are still the center of discussion. Nothing is greater proof of a man's quality and of the living nature of his thought than that it continues to provoke a permanent scandal after his death.

Alexei Matveyev I should like to say something about the problem which was of the greatest concern to Einstein but which has not been discussed here in great detail, except in the introductory speech of Professor de Broglie. I am speaking of the problem of dualism between particles and waves.

Einstein devoted very great efforts to the consideration of this problem because of its great importance for the further development of science. As we all know at present, the special theory of relativity and also the general theory of relativity are very well established theories. Recently a work was published in which the author deduced Einstein's equation of the theory of general relativity from the picture of quantum interaction through the exchange of gravitions. So I think we now have an understanding of the relation between the geometrical picture of Einstein and the picture of interaction which is used in the quantum field theory. On the other hand, the quantum field theory is experiencing major difficulties and everybody is expecting some new ideas which may enable us to overcome these difficulties on our way to the formulation of a unified theory. It seems to me that in considering these difficulties, the problem of dualism between particles and waves is of the greatest importance.

As you will remember, about forty years ago, the consideration of the problem of dualism led Professor de Broglie to a discovery which subsequently enabled Schrödinger to formulate his theory of wave mechanics. So now we are also expecting some new ideas in the field of quantum mechanics which will perhaps enable us to take some further steps of the same kind as those taken in the 1920's.

At present, there is no unified understanding of the nature of this dualism between particles and waves. There are scientists who support the idea that elementary particles behave like waves because they are composed of waves. This idea was put forward by Schrödinger. There are also scientists who support the idea that elementary particles behave like waves because they interact with some hidden field which causes them to behave in a different way. The latter idea is supported by Professor de Broglie, Professor Vigier and others. So this is a field of very interesting work at present. I should like now to express my own ideas in this field.

The aim of theory is to understand, to create a picture in our mind of what is going on in nature. What are the tools of this understanding? The tools of this understanding are our ideas — the ideas which exist in our heads. And what is the origin of these ideas? The origin of these ideas is our experience, and these ideas have been elaborated in the course of the very long history of mankind.

Our experience throughout history was first of all macroscopic experience of small velocities, and therefore we may be sure that these ideas elaborated in the framework of macroscopic experience and small velocities will be applicable only in the same region: that is, to macroscopic experience and to small velocities. Thus if we move outside this experience, we may reasonably expect our ideas and our way of thinking to be no longer applicable. History confirms this. At the moment when we began to study the phenomena characterised by the speed of light — very great velocities which are outside our everyday experience — we came to the conclusion that our way of thinking and our ideas were not applicable to new conditions. As a result of this inapplicability of our old ideas to new conditions, the theory of relativity was established, which changed our understanding of space and time. This demonstrates clearly that when we go outside the realm of applicability of our ideas, we are obliged to change our ideas and way of thinking quite drastically.

Then we went on to study microscopic objects. This too is outside our everyday experience. It is not surprising, therefore, that our old ideas and way of thinking proved to be inapplicable in this new field. However, we are limited to those ideas that we really have. What should we do? We must either go forward or we must stop. There is no other choice. So at this point a concept of dualism of particles and waves was adopted. This concept is simply an expression of the inapplicability of our old ideas and ways of thinking to new conditions. In other words, it is a simple expression of the fact that so-called particles are neither particles nor waves: they are some synthetic entity whose projections on different planes are seen by us both as particles and as waves. Particle and wave concepts are no nearer to a concept of the real entity under consideration than different plane projections of a space object are to the space object itself. Therefore, I consider that we must now recognize this fact and not try to reduce existing reality either to particles or to waves. There is no experimental evidence that such a reduction is possible. The concept of wave package contradicts the experimental fact of stability of so-called particles. The concept of a point particle interacting with a hidden field has no experimental support. Where is the proof that this hidden field exists? The only motive for introducing a hidden field into the picture is the desire to conserve our picture of particles which has been formed during the last two thousand years of the history of the development of mankind. Thus the motive of this development is only the preconceived idea that any new facts may be understood in terms of what mankind is accustomed to. But history has shown that this is not the way in which mankind progresses. From this point of view, I think that an attempt to understand new and complex phenomena in terms of conventional ideas will not succeed. So I feel that at present we must move to create out of these projections we have — particles and waves — a synthetic picture of what we call particles, and this will be a real step forward.

L. de Broglie I am sorry to have to say first of all that I do not agree with Mr. Matveyev. I think it is a perfectly healthy attempt and one which could pay dividends to try to get down to more precise ideas than "a particle spread in the potential state throughout an entire room such as this one". I do not know what this means. If it means it is somewhere but we do not know where, then alright. I can say, "My friend is in Paris; I do not know whether he is in the Champs-Elysées, but he is here somewhere." But what people seem to be saying about the particle is that it is everywhere at once but suddenly condenses into a single point. This is as if I were to say, "My friend is all over Paris then he suddenly condenses in the Champs-Elysées." And this I cannot understand at all. I have made many criticisms of this kind and I have come to the conclusion that the present interpretation must be changed. But I do not wish to embark here on a discussion which will carry me too far, as I want to come back to the question of Einstein, which is what we are here to discuss.

Einstein was, in my opinion, much more positivist than is commonly supposed and much more objective. This was said by Mr. Vigier and also, I believe, by Mr. Ullmo. He looked at things very objectively; when he started his studies on special relativity he had various indications, Michelson's experiment among others, which had put him on the track; he had thought about the physical phenomena, I think, before he considered grand principles. Perhaps much later, when he constructed general relativity, he allowed himself to be guided by the principles derived from special relativity, but I do not believe he did so at the beginning. If you read, for instance, as I have done recently, the demonstrations he gave for the formula of the inertia of energy which plays such an important part in questions of nuclear energy, he takes quite concrete examples, studies them in great detail, as a physicist, and draws from them the principle of the inertia of energy. This was his method of procedure in all the early part of his work which is hardly mentioned today. He analysed the fluctuations and stimulated emissions, which nowadays play an enormous part in the theory of masers and lasers, by very simple procedures which are related to direct experience.

You will see that he did indeed think through all his experiments and this is why, when he discovered the dual nature of light, he at once concluded that there were both waves and particles and that they were intimately linked. This idea permeates all his early writings, and I recently had occasion to read a paper which interested me very much. Mr. Martin Klein — and he is not an accomplice because he is a stranger to me — sent me a paper analysing the complete works of Einstein from the point of view of the matters I have just been discussing. He made a comprehensive study of them and repeated all Einstein's demonstrations. He shows the advance of Einstein's thought in this domain, a highly objective advance, looking into everything very carefully; I made some use of this paper in my remarks at the opening of this session.

He certainly did have a most objective mind and this is why he never really understood and probably never accepted the conceptions of the Copenhagen School. I think, however, that one should never lightly dismiss objections raised by an intellect like Einstein's. Even in the brief introduction he wrote specially for the book presented to me on my sixtieth birthday, he made some comments which are very profound when one really looks into them. Naturally, one has to have thought a great deal about the matter in order to appreciate the profundity, but profound they most certainly are. And he wrote this piece in 1952, only three years before he died.

Reverend François Russo I merely wish to dispose of an ambiguity concerning this idea of positivism. Positivism is a tiresome word because it has several meanings. If we understand by positivism an attitude which not only restricts itself to observation and refuses to recognize any ideas other than those backed by experiment, but is unwilling to stand back and look at the observations from a distance, then Einstein was no positivist, and he moved further and further away from such an attitude. But I do insist that in 1905 his essential procedure stems from a positive attitude, because he did indeed adopt such an attitude when he said in effect, "I cannot accept absolute time because I do not know what it is; I can only accept a time which can be attested by observation." In fact, Einstein fought on two fronts: on the positivist front in 1905 and again in 1921, when he attacked the philosophers who hamper the progress of science by putting the unscaleable heights of the *a priori* before the fundamental concepts of empiricism; then he fought on the front of objective realism when he affirmed that there are realities known to reason which are beyond positive observation.

No, I am not happy about the ambiguity of the word positivism.

I also wish to add a comment on a matter of history. One cannot say that Mach was a positivist in the narrow sense of the word. Mach also knew that it was necessary to stand back and look at experimental evidence. Frank, in his book on Einstein, even quotes Auguste Comte, the Father of Positivism, to the effect that one should be willing to consider realities which are beyond experience.

A. Lichnerowicz I would like to record my agreement with what Mr. de Broglie has just said. I too believe that Einstein was a destroyer of idols in the sense that he would not have any extrinsic philosophy external to his scientific procedure. Any concept which did not pass his double screen of mathematical coherence — which played a very important part in the second half of his life — and experimental analysis, constant observational analysis, was for him suspect and condemned. I think this very rare scientific courage is one of the keys to Einstein's genius. A man who wants to make explicit what is implicit in the philosophy of a given physical theory cannot formulate it other than *a posteriori*, thus it could never be the starting point for a physical theory.

J.-P. Vigier First a comment regarding the School from which Einstein graduated. Mach was a very great physicist; he was perfectly consistent when he denied the existence of atoms on the pretext that no one had ever seen them, at the time when Boltzmann and Maxwell were working out their theory of gases. I think there is a real disagreement between us here and that Einstein was indeed no positivist at heart.

My second comment concerns Professor Matveyev's observations. I do not think that we should see in either Einstein's or Mr. de Broglie's procedure an attempt to go backwards as regards either quantum or classical mechanics. The attempt, on the contrary, is to find new, deeper concepts which would enable them to explain what happens. This is typical of Einstein's approach, to reject a collection of minor rules and look beyond them to a deeper reality. The image of the void which emerges from current quantum theory together with the so-called fluctuations of the zero point and infinite energies, all this suggests that as one descends step by step into the realm of the infinitely small, one finds more and more fabulous and chaotic sources of energy. So there is absolutely no question of reinstating behind the true reality of things the concepts already formed on the basis of another type of experience. But the central problem is to know whether, behind the reality of the chaotic appearance presented by elementary particles at the present time, for instance, there may possibly be a theory able to explain the diversity of the phenomena.

I think we should not see in the last years of Einstein the actions of a man clinging to outworn ideas. That was not his trouble; it was rather his refusal to recognize two sorts of barriers. The first barrier was the postulate that the theories then being proposed to him were the ultimate limit which could be reached by physics. I believe anyone who ever said physics had reached its limit at a certain point in time has been proved wrong by its subsequent development. The second barrier was experimentally based research and research with models designed to represent reality. Both these methods were used by Einstein and ensured the success of theoretical research in 1905; I firmly believe they are just as valid today.

F. le Lionnais Perhaps I might recall — though I except everyone here already knows it — the story of Lord Rayleigh who round about 1898 said that physical science was practically complete. There were just a few constants to be determined and a certain amount of work to be completed, but all the big ideas were already settled and there were just two small details to be cleared up in the next few years: backbody radiation, and the Michelson experiment. He was absolutely right, of course, but the results were not quite what he anticipated.

J. Ullmo Mr. Vigier has just said what I intended to say much better than I could have said it, but I think that the attempts being made at present are not attempts at reduction so much as attempts to work out new and more fundamental structures, able to account for reality on our scale, but without reducing this deep reality to crude macroscopic images. I found the example given by Mr. Matveyev particularly striking. We did indeed have to give up our macroscopic slow-speed habits once we became exposed to the high speeds of special relativity, but this was pure gain to our intuition. My teacher, Langevin — how I like to use this phrase! — often used to say, "The concrete is what used to be abstract before we understood it." Well, we hope soon to introduce new structures and when they are first presented, they will be most abstract — perhaps it will be Mr. Lichnerowicz who will do this at the "leading edge" of mathematics — but then they will gradually become familiar and hence concrete. I find this a comforting thought.

Now that I have mentioned Mr. Lichnerowicz, may I add that he told me that general relativity did not present specifically in an invariant form; but historically this was the idea that guided Einstein, and it is most curious that a universal idea which today inspires all theoretical research — research into invariances, research into what are called fundamental groups underlying objective reality — a method so universal in its application should have been discovered in a singular way, as one might say, thus giving Einstein the impression (no doubt illusory) that by applying this method he would come to understand the entire universe. Far from it! The method was perfect and has become more or less the only method used by modern basic research, but we still do not know what it is that is invariant. In one of my books I have called this the "hypothetical *a priori*". We know we are looking for groups, but we haven't a clue what they are!

F. Gonseth My age gives me a particular advantage which I probably share with no one else here: this is to have been for several terms a pupil of Einstein and to have heard him speak many times. Because I knew him personally, I was able to appreciate his thought in a manner of speaking from inside as well as outside.

Now, my picture of the living, reacting Einstein embraces everything that has been said so far. He *was* a positivist, but he was more than just that; he *was* a theorist, but he was more than just that. I will give you a few brief examples of what I mean.

At a public lecture, which is not much quoted though it made quite a stir at the time, Einstein explained to the members of the Zurich Natural Sciences Society how to approach the theory of general relativity in terms of ordinary experience. The image he used was that of an elevator and, using this example, he explained how a field of gravitation could be suspended; he passed from there to the principle of correspondence. Many times I have heard Einstein expound his way of conceiving the passage from special to general relativity and the principle of correspondence; he stressed the observational rather than the mathematical inseparability of the effects of gravity fields and acceleration fields. This fact of observation was for him of fundamental importance. I am sure all the previous speakers are well aware of this but, all the same, I am a little surprised that they did not emphasize it more, because the way Einstein had of putting a simple principle at the origin of his ideas and taking off from there crops up time and again throughout the whole of his life and thought.

I have told you about his course in classical mechanics; but I could not go into all the details. What impressed us as very young students was his extraordinary capacity for taking a problem and setting it out in such a way that the principles seemed to apply themselves and the solutions to appear automatically.

Einstein was not a positivist in the narrow sense; I have mentioned the elevator and I could have quoted other examples. Einstein the theorist shines through in his reply to that excellent physicist, Weiss, who was at that time also teaching at Zurich, when he offered Einstein the freedom of his library. Einstein fetched a pencil out of his pocket and said, "For the time being, this is my library." You can see that Einstein was not caught between a restricting positivism and a too narrow theoreticism — he was a practical man, practical as regards both experience and theory. When I studied watch and clock making, I realized that the practical man, the man who makes and mends watches, has one foot in the operational and one foot in the rational camp. All scientific practicalities, too, seem to me to be of this kind, and my picture of Einstein — perhaps I may say my global and almost unitary picture of him — is of a powerful intellect of this dual type.

Vladimir Kourganoff I should like to pass a remark on the rather rarefied discussion which has raged around the question whether Einstein was positivist, objectivist, phenomenologist etc. My comments are inspired by a suggestion made by Mr. André Lichnerowicz in an article entitled "La condition humaine du savant". He said that all through the scientific adventure there is implicit an autonomous philosophy which is difficult to pin down but whose pre-occupations, problems and concepts are quite foreign to those of traditional philosophy. It might be a useful social exercise to work out patiently and over a long time the nature of this implicit philosophy through which scientists all over the world are working, researching, thinking and sometimes inventing and discovering.

In other words, what we need in science is a philosophy which has grown up out of the actual scientific method; it is difficult to graft the ideas of traditional philosophy on to problems such as those we are discussing today.

I think we can avoid several misunderstandings and all this rather airy-fairy discussion about whether Einstein was or was not a "positivist" by attempting a direct analysis of his scientific procedure when he created the theory of special relativity. Basically, he proceeded in very much the same way as Newton when he discovered the principle of universal gravitation. Newton renounced all speculative ideas, for instance, Kepler's when he wondered whether the activity responsible for the motions of the planets was magnetic or something else. Newton said to himself something like this: "I do not require to know the *nature* of gravitation. All that interests me is the law it obeys."

I think Einstein proceeded in exactly the same way when he provisionally suspended investigations on the nature of light and the question underlying all the other theories of his time — waves or particles? — and asked himself whether the principle of the independence of the speed of light *in vacuo*, relative to an inertial movement in the reference laboratory and relative to the speed of the source, could explain the experimental results, even at the price of changing our concepts of time.

It was this provisional renunciation which constituted the originality of both Newton and Einstein as against their contemporaries. Einstein relinquished the hypotheses about the propagation of light along with the hypotheses about its nature, in order to concentrate on the study of its properties and their observable consequences.

I should be inclined to call this "positivism," but if the philosophers have a different definition of positivism, I would not presume to put mine forward. What counts is the idea and not the word that stands for it.

However, Professor Matveyev says that "what we are trying to do is to create a picture in our minds of what is going on in Nature"; I incline to the belief that at the time when Einstein was creating his theory of relativity, he was assuredly not trying to do this. He was not in the least concerned with *what* was going on in nature, but *how* it was going on — which is quite a different matter. This difference is absolutely fundamental.

Olivier Costa de Beauregard Among the many comments I should like to make, I will select just two brief ones.

The first is that in our lifetime we have witnessed a great event; the reconciliation of two of the great twentieth century theories — relativity and the quantum theory.

On the plane of special relativity, this year's Nobel prize has crowned the three great architects of its success: Tomonaga, Schwinger and Feynman; I should also like to acknowledge the remarkable work of Mr. Lichnerowicz who has done the same for general relativity and the quantification of the field of gravitation.

My second comment concerns the very great interest which I think lies in a new series of experiments, designed to reveal the physical role of the potential electromagnetic vector in a region where there is no field; I think this is important from the unitary point of view. The most recent experiment of this kind is that of Mercereau and his co-workers who, in agreement with earlier experiments, have shown that the electromagnetic potential exerts a physical action where no field exists. The result is invariant of gauge, but I think the most reasonable conclusion is that physically there is no invariance of gauge, for how could one conceive that a magnitude which has a physical action could be defined within a wide arbitrary approximation?

Now there is a theory which can account for such a situation, that of the photon of non-zero mass which Mr. de Broglie has always defended and which I, too, have always supported and still do, particularly since these new experiments. I think these experiments have shown that the photon is more entitled to its place amongst the other particles than people had been willing to believe before. This concludes my second comment.

F. le Lionnais Before I close this debate, I would like, with your permission, to relax the atmosphere a little by recalling another aspect of Einstein's scientific activity. He did not exercise it solely upon the mountain peaks of science but in the valley bottoms, too, if I may so express it.

This very great man was passionately interested — mostly in his younger days, it is true — in quite modest theories and practical inventions, a reminder that he started his career in the Patents Office in Berne. I hope you will not take it amiss, but I have selected two examples from among a dozen or so with the aim, not just of amusing you, but also to show you there was another side to Einstein.

When he was young, he devised a theory to account for the bends or meanderings of rivers, and he illustrated his explanation with an experiment which can be repeated in a teacup and enlarged to rediscover Baer's geographical law, according to which the waters of the northern hemisphere, the one we live in, tend preferentially to scour the right bank while the opposite phenomenon, of course, occurs in the southern hemisphere.

He also took an interest in the law of the mechanics of fluids, formulated by Magnus and developed by Prandtl, the law which was subsequently applied to propel ships by means of screws instead of sails.

It would be possible to give you many more such examples to show that the maestro of fundamental science did not despise practical applications.

I shall not venture to sum up this debate; its richness is a reflection of the richness of Einstein's work and of the diversity of those taking part. For, while we are here under the sign of synthesis, we are not under that of syncretism and our thoughts need not necessarily coincide. I do think, however, that one thing that emerges from this debate is that Einstein was initially very naive — happily for us — in his scientific approach, yet he was at the same time revolutionary and non-conformist. I see no contradiction here. I think that to assert naivety aggressively against traditionalism is one way of being revolutionary. His contemporaries — Poincaré has been mentioned among others — were far from naive. But he had the courage of his naivety and this is perhaps the moment to quote Einstein: "the true scientist must be either in love or a believer".

Towards a Cosmology

François le Lionnais I call upon Mr. René Poirier to open this debate and to explain its particular interest for philosophers.

René Poirier I had better explain why they asked a philosopher to open this session: for many centuries cosmology was a speciality of the philosophers, and rather a despised one at that, whether it was dogmatic, by which I mean a doctrine concerning the form and origin of the universe, or critical, by which I mean a discussion of the contradictions or even the illusions of cosmology. People saw it as unbridled and unfounded speculation, sheer prejudice, or local intuition raised to the status of natural and intellectual laws. In short, it was rather a disreputable speciality. As for the learned men, they were at most able to show the difficulties inherent in the laws of force or illumination which, when imprudently applied, brought in the inverse square of infinite space, which was supposed to be peopled in a homogeneous manner; or again, they asked themselves questions about the evolution of the stars or the origin of the solar system — and these are still local problems, albeit on a huge scale.

Today, it is the other way around and cosmological speculation is based essentially on mathematical systems; these are inventions of the mathematical spirit, tangential to local experience, which would easily impose themselves on us if only — and perhaps, I may add, happily — the legislators who wish to give us this categorical system of law were not so often in disagreement with one another. In any case, the philosophers are requested to keep out, not to impose conditions on the systems and even to refrain from suggesting possible explanations for them.

I am here to put to you the view that this attitude is rather exaggerated and that cosmological hypotheses have to be interpreted and even selected within a context which can never be either purely experimental or ruled by logic and mathematics. For instance, I should like to ask what cosmological problems relativistic theory creates for philosophers; by cosmological I mean relating to the entire structure of the world, the physical universe, not only its size and shape. I shall try to guide you gradually towards my conclusion, which is, without more ado, this: I think that even today it is difficult to construct a cosmology without raising questions of ontology and tacitly assuming an underlying methaphysics. As soon as cosmology tries to assume any sort of reality, it must conform to certain limit conditions and these, though not in themselves extra-scientific, lead back to problems which are of a different kind from those it started out from. I mean problems like the relationship between body and soul, the possibility of freedom, and so on.

So I must devote some time to epistemological problems.

The world is not simply what it is; it is also all that has been built up by discussion conducted in accordance with rational norms. I mean, of course, the world in its entirety; for local science there is a radical and definitive criterion — experience. On the other hand, when it comes to vast extrapolations, the logical mind is very willing — almost too willing — to construct them. However, cosmology, like art, lives by constraint and dies of too much freedom.

We must thus consider what rules we can impose on the cosmological discussion and, because we are looking at it from the relativistic angle, this forces us to examine the classical problems of interpretation.

First of all, there is the problem of the geometrical explanation, which has already been alluded to. Is this a *ne plus ultra?* The way in which special relativity, for example, considers phenomena within their own systems and subsequently transcribes them into local systems is a trick which has been remarkably successful, and we do not require to know what goes on between the two types of systems in order to make it work. It is a self-sufficient system of transformations, just as Newton's physics of remote actions was self-sufficient, simply because it did not seek to find out what gravitation was and how it was propagated. One could say the same within the frame of general relativity about the deformation of space and time in the proximity of a gravific mass.

Is this a definitive, final form of science or should we, on the contrary, look for an explanation beyond this, an explanation of a type yet to be defined but which we should like to be both fairly concrete and fairly intellectual? Basically this problem has a certain symmetry in common with the problem we are to discuss here, regarding the indeterministic and probabilistic interpretations of quantum mechanics. Are these the ultimate, or can we imagine an infra-determinism which would justify them, though of course the conditions of explanation would have to be different from those we are used to now?

This ties in with the fundamental problem which ought to guide all our thinking on cosmology — indeed, all our thinking of any kind, any epistemology. There are two terms to which we hold firmly: experience on the one hand — not always perfectly defined, but still we know in broad outline what sort of facts are acceptable as paradigms and ultimate norms for any theory —

and on the other hand logical consistency, which is basically the only norm for mathematical invention. Experience and logic — but can we claim that all science is reducible to logically coherent discussion resting locally upon experience? Or do we have to admit that there is some missing link between them? On the one hand, we see and describe the world and, on the other, we discourse of it according to the rules of a logical language. However, there may be another way of thinking about it, lying somewhere between the two, and this need not be a rehash of concepts and intuitions borrowed from our present range, though it may still have something intuitive about it.

Mr. Matveyev referred in the previous debate to new forms of thought, brand new concepts, and stressed the importance — to borrow a metaphor from the Gospel — of not putting new wine in old bottles. He spoke of the need to re-educate our minds, refashion our intelligence and develop a new kind of awareness, better adapted to the new requirements.

Now, this is a very tricky problem. Can we slough off our skins like a snake? Is it really possible to remodel the human mind? We can, of course, change its models, find out more about the way it works and cut its pretensions down to size, for example, by applying the process of demystification (the word used by Jacques Merleau-Ponty in his thesis, *Twentieth-Century Cosmology)* to a number of principles hitherto considered as absolute, a typical example being the law of the conservation of energy. Clearly, all these principles are mere habits of thought, often confirmed by experience and generalized by induction, but all the same not necessarily representative of the human mind. We are frightened to use our freedom, we seek new contraints because we cannot imagine this world without them; however, this is not what I want to talk about.

The last epistemological problem is this: is it legitimate to invoke something beyond experience, even though it cannot be determined? In other words, should we submit to a strict positivism which not only says we should ignore metaphysical pseudo-edicts (that goes without saying!), but also tells us that nothing exists for the scientist beyond measurable, concrete experience and that anything else is just pseudo-reality? Or should we believe, on the contrary, that beyond the things that can be subjected to direct experience (though, in point of fact, we never know where to draw the line), we must inevitably refer back to a reality which, though it has some affinities with the world of science, yet differs from it in depth and dimensions, and which, taken as a whole, constitutes relative to science a sort of semi-transcendental Beyond?

But, you will say, isn't this just a Spencerian unknowable? No, I mean something else, a Beyond which can reasonably be invoked to explain the sum total of experience. It must be able to explain the world of physical theory, yet meet certain philosophical requirements, such as making more intelligible problems like the existence of a sensible world which is the object of our awareness, the existence of the human spirit and possibly, too, the existence of freedom.

These, therefore, at a first approximation, seem to me to be the essential limits — a mere outline of themes — for an absolute operationalism and a science which would admit as a criterion and means of verification only local convergence with genuinely experimental results. If you like, this is the problem of research on absolutes, in the sense that it is reasonable to assume some things in advance and reasonable to attribute certain properties to them, although strictly speaking it is impossible to prove them.

This brings me now from epistemological considerations to more concrete problems.

These cosmological problems comprise, first of all, in the extension of the theory of relativity, those concerned with space and time. One of these is absolutely vital and I propose it at the start because it is one of the sort which comes up against limitations which, though not exactly extra-scientific, are metascientific at the very least. Should the space-time continuum in its indissoluble unity (there being no free and natural distinction between space and time) be considered as a dogmatic implication of relativistic theory (this is Mr. Costa de Beauregard's view, and I do not think I am giving away any secrets in telling you this), or should it be seen as a stratagem of mathematical exposition? The non-existence of a universal present, of a cosmological time, should perhaps be the only thing to be regarded as legitimate, since the various types of time and space are to some extent ideal sections of this continuum.

Ought we to take literally the existence of a single, fundamentally indivisible reality of space and time, or should we admit that this is basically just a procedure used in exposition, infinitely valuable, of course, but no different from the other matters I was talking about earlier — geometrical optics or the physics of remote action; should we also admit that, if we try to adapt discussion to whatever the reality of things may be, it is better to accept the existence of a separate time and space.

This is a very profound question because, if we admit their fundamental non-distinctness, and thus the fundamental unity of the space-time continuum, then not only is the past preserved in all its spatial-temporal extension but the future, too, is in a sense there, and the process of becoming is merely the advance of pure consciousness moving across the space-time continuum along lines of increasing entropy.

Here is a first problem which we shall perhaps discuss later on. Going further, there is the problem of realistic interpretation of the multiplicity of times and spaces in special relativity. True, everybody is agreed that experimentally, that is, physically, all these times are equally real; there is no question of contrasting an allotted time with a possessed time, or a fictitious time with a real time. It is quite clear that all the measurements of space-time made within the various systems in uniform relative motion are equally real. Everyone also agrees that there is no way of attributing a privilege to any one of them so long as we adhere to the systems in relative inertia; the same, however, would not be true of systems in relative acceleration, most relativists being apparently willing to admit that accelerations are absolute in character.

Are we consequently forced to relinquish the old Newtonian idea that there is basically an immovable, corresponding to the existence of a substrate, incidentally perfectly indertermined and indeterminable, but endowed with a sort of substantiality, a sort of common present, and consequently having a spatial dimension and a temporal dimension of events at once separate and well-determined, although none of the measurements that we can make of them enable us to say what they are? This is tantamount to saying: given that the theory of special relativity generalizes in the ensemble of electromagnetic phenomena the properties of relativity which Newton's remote actions possessed (he, by the way, inferred from them not the non-existence but rather the existence of absolute space and time), given this, could we, or should we, continue to be Newtonians? And if not, what sort of idealism should we adopt and how can we effectively imagine this plurality of space and time?

This question links up with the famous exclusion of the ether — a dirty word nowadays, but one which is for ever cropping up again in various forms as a support for all the proliferating fields, the background from which we cannot escape.

What I want to say is simply this: that to the extent that we try to propose a problem of objective reality, of the onology of time or space and of the physical world in general, we are forced to accept what I would call the cartographic or radiographic conception of science.

I will illustrate this briefly. Let us admit that the sensual world, the world of experience with its colors, its solidity, its measurable dimensions and its actions, is in relation to reality what the visible form of the human body is to the actual body. Under these conditions what science presents to us is not all the actual human body but an X-ray picture of this body or, better perhaps, a map, also made from the outside but more penetrating, more geological; when I call it an X-ray, I mean not only that it is more useful — because an X-ray helps the surgeon more than a photograph — but also that it is more faithful, since it shows the structure of the human body in depth in a way that the visible image does not. However, no one would dream of asserting that X-rays of the human body *are* that body and hence the true object of our perception. No, there is something else, something beyond, which is expressed in two languages — a sensual and an intellectual language, both of which are "superficial". (This intellectual language is, moreover, a presumed language, since in nine cases out of ten we do not even make the X-ray pictures, we just imagine them; but all the same, we sometimes really make them).

If this is the case, the conditions which can be imposed upon a cosmological theory should take into account the fact that the being in question is not homogeneous, for instance, with this three-dimensional world we live in; it is something beyond, of which this world is the projection, a mere reflection.

I now come to my third and last point which corresponds to cosmology in the narrow sense: it is the principle of intellectual totalization of the data of experience. I must confess we clothe in words a world which we know locally through experience, or, to put it more simply, we construct grand abstract systems which are purely mathematical and we make them fit the facts of our experience here and there; these systems are at the same time a totalization. But to what extent do they totalize the real facts in the full sense of the word? What do they really represent? What is it we are totalizing? Are these mathematical constructions, logical stratagems or a true expression of something real?

Here again, I had better explain what I mean by a comparison. When we look at the heavens, we see the stars, we see an astronomy of position and a geometry of the celestial sphere. It is perfectly reasonable to say that, locally, the geometry of the heavens is of a more or less spherical kind. But if we integrate and then say that the sky as a whole is shaped like a vault, does this still have any meaning? We have totalized our experience and yet — *what* have we totalized? Something which does not exist! We have totalized, or integrated a manner of speaking; note, moreover, that we shall sometimes have to say that in reality the heavens are not shaped like a true sphere but more like a flattened sphere. And yet we know that this statement is meaningless; if there were a celestial vault, it would be spherical. But there isn't one! The problem for those who practise cosmology is, as I see it, rather of this order: relative to reality, is the whole of this experimental, three-dimensional universe a perspective view, a projection or a simple sensory reflection? And when we make bold to integrate it, do we achieve anything more than a mere verbal integration, a purely mathematical scheme? Can this be anything which genuinely exists?

Must we conclude that this reality is an unknowable? I made a reservation just now about the right to speak of things whose existence we are obliged to accept but whose essence, as the pedants would call it, cannot be exactly determined.

At a time when, as Mr. Lichnerowicz succinctly pointed out, there is now from the point of view of ideas a sort of preponderance of the global over the local, the fundamental problem of cosmology is to identify the variables. What do they correspond to, either in relation to a possible experience or in relation to reality? The reply is difficult indeed. Sometimes we rashly attempt to correlate things experienced with the totality of values of cosmological variables. It is most instructive to see Gödel, writing of "Einstein as philosopher and scientist", saying of certain singular lines of the time order, which would form a closed loop; they cannot correspond to any reality because psychologically this would mean that we relive states which are both determined in the past and free in the future. This is a very subtle matter; I do not think one can defend this position for one second — the problem is more more complex than this. I quote this example merely to show how we unconsciously apply psychology and metaphysics the moment we try to identify and interpret purely abstract variables in terms of experience. Without going as far as this, we can ask what experiences of a purely physical type could we, in fact, get to correspond even locally? Dare I utilize an aggressive term and say that to accept a cosmology is *a priori* the very negation of any principle of operationalism? We are quite well aware that, under the conditions of cosmology, which has origins so extremely remote that the world as it was then had little or no resemblance to what we see today and where we cannot conceive of a clock in our meaning of the word, if we still wish to give a physical meaning to time, it can only be the measurements that would be made — to use Huyghens' expression — by a cosmostheoros. Is it physically possible to identify variables in the more distant regions from the results of experiments which will be anything more than fictitious?

The problem — and I shall stop here — becomes even more painful and difficult the moment we try to make use of these cosmological considerations to resolve that most fundamental of questions: the irreversibility of time. To what extent can a cosmological hypothesis enable us to escape from the antinomy of the irreversibility of time and all that this implies concerning — not the end, perhaps, since we seem quite prepared to accept that the world will come to an end one day — but the beginning, at any rate? Many people have great difficulty in believing that it ever had a beginning.

For a long time attempts were made to adapt physical hypotheses to the cosmology of irreversibility, a sort of pseudo-philosophical quest for the endless return. This endless return, in the form of a reversal of time (the universe reversing back to its primitive origins, or perhaps even oscillating eternally) or of a time cycle (like a book written on a cylinder whose end joins up with its beginning), is something we cannot believe in. This derives from sheer prejudice, which I will call anti-theological. But then again, the alternative hypotheses are equally unsettling. We are shocked by the idea that the world had a once-only absolute beginning, and even the biologically inspired hypothesis that the world will decay like a human body, leaving seeds from which a new world could spring up, perhaps embodying something of the old one in its make-up and carrying it on in the same way that life is carried on from one generation to the next (and this is perhaps the most natural and least repellant hypothesis), has not won unqualified support and we do not presume to impose it upon everyone.

I shall stop here. I hope I have shown without causing too much offence that there is a certain solidarity between scientific cosmological hypotheses and philosophical cosmology in all its non-autonomy and non-dogmatism.

A. Lichnerowicz I should like us to look upon the problem of cosmology as a scientific one, and in this sense I am a disciple of Einstein. First, I wish to record on the one hand my profound agreement with my old accomplice in cosmology, Professor Poirier, and on the other hand my disagreement with him on certain points.

First, I shall try to talk to you about cosmology in the narrow — indeed, the narrowest — sense of the word. I refer to that cosmology which one might say came into being in its general conception with Einstein. What is its objective? It does not claim to be an integration of all phenomena (this is why I call it cosmology in the narrowest sense), it postulates a world in which each galaxy is seen as a point. It is located on a certain scale, at a certain order of size, and its cosmological ambitions do not aim to account for little local details but to see things on the grandest possible scale.

I shall affirm right from the start — in common, I believe, with Professor Poirier — that anyone who takes science seriously does not accept scientific cosmology as science. It is a poem to science, a game with science, an ambition of science, but emphatically not an integrated part of science; there is no cosmology today which can be called scientific in the sense that it offers great scientific theories.

I think this is important; I believe that this distinction is not made sufficiently clear in the literature — and this includes some very serious books indeed. One gains the impression that each thinker has his own pocket cosmos tucked away somewhere, competing with all the others, and that this cosmos must be taken just as seriously as the rest. They are indeed fascinating poems and games. What makes them so fascinating? Each of them inspires new types of observation and experience and coordinates them in some way. This is a powerful intellectual motor for a whole field of science, but it lies outside the rigor and the earnest strivings of the true scientific adventure. We need such games in order to hearten us for the often austere daily round which enables us to weave, day by day, innumerable small facts into the true fabric of science.

How was modern cosmology born? The first attempt was made by Einstein himself with a static spherical universe. It is readily seen that such a universe has the great drawback of being unstable, and we are not inclined to accept a universe which can be thrown into a state of motion by the slightest random disturbance.

Then there were the enthralling enterprises of Eddington and, above all (and also the most popular to date), the universe of Friedman and Lemaître, the expanding universe. The interest and success of this model stem from the fact that it supplies an approach for integrating observations on the red shift and on what has been called the flight of the nebulae, at speeds of the order of magnitude of the speed of light for the most distant ones. In fact, the cosmological problem within the relativistic frame, even at its strictest, is as follows: we have field equations, which are local. We have to find in space-time, whose topology is unknown, solutions which will be global. We have to ask ourselves: what reasonable statement can we make about the topology of space-time?

In point of fact, we cannot say much at all. It is all a matter of taste. We have to consult the philosopher about which cosmologies would meet the requirements from our double point of view and to find, if I may put it this way, universes which will be congenial. Generally speaking, we find that those with a closed space are the most congenial as they eliminate a great many difficulties.

Another and more subtle problem which is of extremely great interest is that of orientability. Is it conceivable that, if we made a long enough journey through our universe, we should come back to the same event but with the symmetry reversed? This is a property which, while by no means excluded, is in the general coordination of experiences and facts relatively "uncongenial"; the explanations in terms of Lorentz and Poincaré groups, the existence of two types of neutrinos together with their anti-neutrinos, one neutrino being linked with the muon and the other with the electron, would lead us rather to suppose that the universe is orientable.

The problem of time arises, I believe, under two fundamental aspects. The first is the orientation of time, the arrow of time; what in our universe authorizes us to speak of past and future, and how can we distinguish the past from the future? This is a first fundamental problem for the study of a topology. A second is the topology of time itself. Professor Poirier mentioned Gödel's universe, a universe of endless return, having lines of time which form closed loops. This does not really appeal to us. Indeed, for reasons which are both philosophical and mathematical, we much prefer time to be open-ended, to extend from minus infinity (zero) to plus infinity.

Zero time, in fact, does not find much acceptance with most mathematicians. This global time should be sharply distinguished from local, measured time, the time we hang by means of our clocks on to benchmarks in space and time. I should like, however, to note a few points on which I disagree with my friend Poirier. I do not believe that the present state of science is a definitive and completed form of science, but I am prepared to wager that science will not achieve any more concrete type of explanation. I think that, on the contrary, it will move from year to year towards greater abstraction. Indeed, what we wish to know in a precise manner within a cosmological framework are, amidst the inexhaustible variety of Einstein's equations, simple, abstract general principles which will enable us to select a universe which will be the best possible model of our Universe. I think we can perfectly well discuss this in non-mathematical language; there is a mathematical language and there is a completely different type of language — that for which Professor Poirier was pleading autonomy, or at least legitimacy. I am willing to grant it legitimacy, but on condition that we realize exactly what we are doing. For there is no question of being objective if we use this type of language. Objectivity is obtained only at the price of harsh scientific discipline, scientific asceticism, and, when we do come to compare our choices, our tastes or our preferences, we shall indeed be speaking of the sensible world but there will be nothing objective about the subject-matter. We shall be in the region of what I may call the necessary and fertile misunderstanding.

One further comment: electromagnetism was said to be in line with Newtonian dynamics. I believe the reverse to be nearer the truth. Electromagnetism received the statute and even imposed its group on Newtonian dynamics; it is this imposition which creates the problem of "unknowable" space and time, and on this I reserve my opinion.

O. Costa de Beauregard I am no specialist in relativistic cosmology. I speak here rather as a cosmologist in the wider sense and as author of two books, the first of which, entitled *Le premier principe de la science du temps, équivalence avec l'espace,* works out the consequences of special relativity in the various branches of modern physics. I am not going to say much about this. A great deal has already been said about the relativistic equivalence between space and time in the C relationship. It is a quadratic equivalence which, in my opinion, is reminiscent of the linear equivalence instituted by Joule's principle between work and heat.

The only point I want to draw attention to is that the application of relativistic covariance in the limited sense to quantum mechanics marked a considerable advance. Each time relativistic covariance was re-established in quantum mechanics, not only was great conceptual progress made, but great progress was also made in the order of numerical explanation. This is true, not only of Sommerfeld's earlier electron theory, but even more of Dirac's electron theory, which was the first satisfactory explanation of the fine structure of the hydrogen spectrum. And it is true, too, of the third stage, quantum field theory, in its covariant form with an explanation of corrections for radiation. We owe this theory to Tomonaga, Schwinger and Feynman, who received the Nobel prize for this work. It is concerned with some of the effects which are predicted with very great accuracy in theoretical physics. (The Lamb-Rutherford effect and the abnormal magnetic moment of the electron).

I shall have rather more to say about my second book, *Le second principe de la science du temps,* which deals with the problems of irreversibility. These problems, which are immensely subtle, have enjoyed a great revival of interest over the past few years. First there was Reichenbach's celebrated book, *The Direction of Time.* After the publication of my second book, I got into correspondence with many people in all parts of the world and we now form a select international club concerned with the irreversibility of time.

When we look closely at the problem of irreversibility, whether in phenomenological thermodynamics or in statistical mechanics, whether in wave theory or even in cybernetics, we see every time that irreversibility is never present in the elementary equations of evolution but is imposed from without, like a boundary condition.

Let us take, for example, the classical calculation of probabilities. How is it that blind statistical prediction performs so well, but blind statistical retrodiction simply will not work? The reply to the calculation of probabilities is, in fact, an appeal to the principle according to which, when one has to deal with the problem of retrodiction, one is not able to regard all possibilities as equally probable, but has to allot to each an externally chosen statistical weighting; however, nothing definitely imposes the choice of such constants: the only statement contained in the principle is that they are not all equal.

If we seek the expectations which motivate this law, we must do so by inserting the system under study within a larger system. Take, for example, a drop of ink which dissolves in a glass of water. What is there to prove that if you waited long enough the drop of ink would not reform in the glass of water? The answer is that the system ink/glass-of-water is in interaction with the rest of the universe: the concentrated drop of ink did not form spontaneously by the evolution of a system, but was dropped from a pipette by a human agent. Thus we have referred our system to a larger one: the man with the pipette. Once one starts along this road, it is impossible to stop; one is drawn on step by step to ever larger systems, and this is why this type of problem immediately acquires a cosmological significance in the widest sense.

Many people think there is a relation between this sort of cosmology, which is statistically achieved, and relativistic cosmology in the strict sense. I am among those who think that the subsequent course of history will show that there is a close connection between cosmological phenomena, properly speaking, and statistical phenomena. In particular, I, like many others, am much impressed by the fact of the reddening of photons from distant sources which also, incidentally, distinguishes the degradation of thermodynamic energy. There have already been many speculations about this, but I think I may say that none of them is definitive. Here is a path to be explored and from which we are entitled to expect interesting developments.

When we look at the universe from a statistical angle, we are not merely referred back to cosmology, but to psychology, too. Nothing, as you are well aware, is as ambiguous as the notion of entropy, or probability.

Let us talk about entropy: it is the logarithm of probability. Ought we to call it objective, or subjective? This is a question nobody has ever been able to answer. There are arguments for and against; I have finally come to the conclusion that we have to say it is both at once. Probability is a field where we capture the interaction between matter and psychism all alive.

When we speak of probability, we have to include a new concept which has been introduced with extremely fruitful results — the concept of information, also defined as the logarithm of probability. Now that it has been introduced, we perceive that this concept of information was always implicit in problems of the calculation of probabilities, such as turning up a card from a stack etc. What do we gain by turning the card over? Information! And I believe that the exposition of these ancient problems by Pascal, Fermat and all those who came after them would gain a great deal if the concept of information were introduced right from the start. So here cybernetics has made a major and far-reaching discovery, as was stressed by Brillouin particularly, viz. that when information is obtained from a physical observation, there is a corresponding decrease in the entropy of the universe. (More generally, this is equally true of any piece of information, however acquired, whether in the course of a telephone conversation or by operating an adding machine.) This is certainly a considerable discovery. It means that we are caught up in a cosmos which is essentially probabilistic, and that we are all, whether we live in this cosmos or carry out experiments in it, in the situation of the gambler playing cards or throwing dice. And, be it said in parenthesis, I think that all the discussions about quantum mechanics would be invigorated by being repeated in a cybernetic context. After all, what do we obtain in a microphysical experiment involving a shift of probability, unless it is information?

There is a certain symmetry about this problem. Cyberneticists soon realized that to acquire information is to become capable of re-ordering the system from a macroscopic point of view. This is what is done by Maxwell's daemon, who has a very detailed knowledge of the speeds and positions of molecules. He possesses more detailed information than that available in classical macroscopic thermodynamics and, thanks to this, he can restore order to the system from the point of view of macroscopic thermodynamics by sorting the molecules and expediting them in bundles.

Thus, cybernetics has brought out a relationship between two aspects of information: information-gained-from-knowledge and information-conferring-power-of-organization. This is an unpremeditated rediscovery of the old Aristotelian relationship between information-knowledge and information-power-of-organization. I shall not develop this point any further, but I think that if I did, it would set us off in directions proper to Bergson and Teilhard.

Jacques Merleau-Ponty First, I would like to express my disagreement with one of Mr. Lichnerowicz' statements. I think when he says there is no scientific cosmology today, he is making something of a value judgement rather than a real one. Indeed, Mr. Lichnerowicz himself added that cosmological theories are poems, but poems which first act as powerful intellectual motors and then enable observations to be coordinated and even stimulate fresh observations. I therefore ask Mr. Lichnerowicz what a theory has to have before he considers it scientific? Is it not enough for it to coordinate existing observations and stimulate fresh ones and yet again, more generally, that it should act as an intellectual motor?

Moreover, these cosmological theories may be up to a point compared with the results of observations. At present remote observation of the universe enables us not, it is true, to chop and choose between the various cosmological models which have been advanced, but at least to eliminate a certain number of possibilities (with a certain degree of probability) and similarly to confirm that cosmological theories should at least obey certain general principles. In particular, the fact that the spectral displacement of the galaxies obeys a familiar law, a truly universal law whose workings can be verified even upon galaxies several thousand-million light-years away (although, of course, with the degree of uncertainty such a distance implies), and that, for example, we obtain the same result on certain galaxies by means of different methods, due allowance being made for errors of observation, all this should be enough to prove that cosmology is scientific. If this is not so, then the word scientific is being limited in a way that I would consider quite excessive.

Furthermore, Mr. Lichnerowicz has every right to call himself a disciple of Einstein, but I would just like to point out that he is not completely faithful to his master, since Einstein himself in some well-known papers considered the cosmological problem to be an important one which ought to be treated within the framework of the theory of general relativity.

Another point I want to bring up is one of the most important questions from the point of view of modern cosmology, for philosophy as well as science. It is a matter of what is called in cosmology the hypothesis of cosmic time. Everybody knows that the theory of special relativity caused the traditional idea of universal space-independent time to be discarded. I believe, too, that another thought-provoking fact is that Einstein himself, wishing to transform the theory of relativity into a cosmological theory, was led, not to re-establish Newtonian universal time, which would have seemed natural, but to introduce a hypothesis which is difficult to express but which involves postulating a universal time standard for the whole world. Einstein introduced this hypothesis under rather suspect conditions, because at the time he did so some observational data, which have been acquired since, were not available. The hypothesis of cosmic time has since been confirmed by observation, first by discoveries of the distribution of the images of galaxies on the celestial sphere, and even more by the discovery of the spectral displacement of the galaxies. And then, there was also — and I think this is interesting from the viewpoint of natural philosophy and not that of science alone — a sort of convergence within this theory of cosmic time of the results of new observations and some theoretical research mentioned earlier by Mr. Lichnerowicz, and these have shown that without this hypothesis of cosmic time, we run a distinct risk of having to admit that we live in a paradoxical universe.

We may, indeed, be none the worse for living in a paradoxical universe; but one is nevertheless justified in asking whether there is not more intellectual satisfaction in living in one which is not.

I am referring to Kurt Gödel's famous model of the universe, which is interesting not only in its results but for the reasons which prompted its construction. We know quite well what these reasons were, Gödel does not attempt to hide them in the short article quoted earlier by Mr. Poirier. Gödel constructed this model of the universe for philosophical reasons. He tells us, "For myself, I do not believe time is real." By this we may understand: "I do not believe time to be something which belongs to the real structure of things, and I will prove it. I shall prove it by showing that Einstein's equations, including the cosmological term, are capable of solution (it is thus a cosmology in perfect conformity with Einstein's intentions) by making time not a wholly closed loop but a set of co-existing open-ended and closed lines of the time order." In this model of Gödel's, as in more orthodox models, the fundamental substrate of nature is made up of open-ended lines of the order of time.

At each point one can define a standard of rest, that is to say, particles which move along these lines; but there are also in this universe closed loops of time, and Gödel even calculates the amount of fuel which a cosmonaut would consume in traversing them. It is such a huge amount that in practice it would be impossible to make such a journey. But, if one could do so, it would mean that in a universe where time as a whole has a unidirectional flow a traveller could take a turn through his own past, and this is certainly very odd indeed!

Perhaps Mr. Poirier is right to say that Gödel himself in the explanations he gives is not explicit enough about the absurdity involved in this return to the past. But at the very least there is something rather surprising about this which makes it unlikely that we could consider Gödel's universe coherent.

I should like to add that there is something rather satisfying and at the same time a little surprising in the fact that the data obtained from the observations of modern astronomers, insofar as they authenticate the existence of cosmic time, protect us from the absurdity which would result from a universe of the type proposed by Gödel. This is a circumstance to be pondered, to be meditated upon; it is in one sense too happy a circumstance not to give rise to the suspicion that it may not be fortuitous, in other words that our own reason is so made that we project upon the world the appearance of cosmic time, or again that we see in the world precisely those things which preserve us from this absurdity. But I do not think that the suspicion is justified in this case, because I really cannot see any secret link which might exist between these subtilizations and the highly sophisticated hypotheses of relativistic cosmology on the one hand, and real observational data on the other.

Let us, anyway, salute their meeting, which inspires a certain optimism; here is a kindness on the part of nature, and modern physics is not accustomed to such things, witness the incredible conceptual and mathematical acrobatics that physicists have to perform when they try to interpret the experimental data of microphysics in a coherent manner.

Now, however, we must temper this optimism because, as Mr. Poirier said indirectly, the existence of cosmic time creates all sorts of highly complex questions, particularly as regards time past. The inference suggested by the spectral displacement of the galaxies leads us straight to an expanding model of the universe. So what happened to start this expansion? And this takes us right back to the metaphysical problems mentioned by Mr. Poirier a time back; however, I do not think I had better go into this here.

Guiseppe Cocconi I want to make two remarks about the cosmological problem that, though down-to-earth, as they express the point of view of an experimentalist, are perhaps not completely irrelevant.

a) I think that one must be careful about giving to relativistic cosmology a high weight, even from the point of view of mathematical or philosophical content; I believe in fact that the main results of the cosmologies we are presently discussing are consequences, not so much of the mathematics used, but rather of the cosmological principles on which the cosmologies are based.

When the problem of cosmology is faced, one generally starts by postulating two so-called cosmological principles: one is that of a uniform substratum and the second is that of isotropy. Once you have accepted these two principles and built a cosmology incorporating them, you are bound to find a cosmology which is essentially the relativistic cosmology. In fact, it is well known that, even using classical Newtonian mechanics and these two principles, one inevitably obtains the main result, namely the expansion of the universe. So I think that though the general relativity theory is the right tool today for handling problems of cosmology, it is only a tool, it doesn't necessarily contain the truth. The final results are already contained in the cosmological principles.

To show the inadequacy of general relativity I have only to remind you that there are still problems, like the problem of inertia, (I refer essentially to the point of view of Mach) which are not at all understood in general relativity and are left open.

b) As an experimentalist, I am puzzled by the following observations. If I look around at the world I am impressed by the fact that the farthest galaxies we can see now, which are five or six billion light years away (following the law of recession), look very similar to the galaxies which are much nearer to us; in other words, I observe a very slow (if any) evolution in the cosmos.

On the other hand, if I look at the world of life, what impresses me most is its strong, fast evolution, especially if I focus my attention on intelligent life.

As an experimentalist, these observations lead me to ask whether the fact that life is expanding so fast, in comparison with a cosmos which is essentially static, could not mean that eventually life will have a strong impact, a cosmological impact, on the whole universe and not remain a phenomenon localized on the surface of the small grain of sand that is the earth.

In conclusion, I would say that I agree with Mr. Lichnerowicz when he says that cosmology thus far is essentially a game, a very intelligent game but not much more than a game.

Stamatia Mavridès I should like to stay on the prosaic ground of present-day cosmology, ground which is essentially close to experience, and I would like to take up a few points suggested by the previous speakers and also to mention some things which I consider important but which nobody has mentioned yet.

First, I want to stress the constrained nature of cosmology since Einstein. Since 1917 the cosmological problem has forced itself in an ineluctable manner on the scientific world. Whereas it had previously been either the fruit of philosophical speculations or, when the speculations were scientific, more or less gratuitous, after Einstein they entered a compulsory phase. Given that general relativity seeks to determine structures linked with a certain material distribution, we are forced to search with or without success for the probable structure of the material content of the universe around us. Thus, in our time, the cosmological problem assumes quite a different complexion.

In the nineteenth century this cosmological problem was more or less despised as being mere philosophical speculation; it was also very much despised by physicists, in the 'thirties, say, because physicists had to use very simple models in order to tackle this problem initially. They considered the exercise of moving from metaphysical speculations to such simplistic representations of the universe altogether too elementary. But these models were inherent in the mathematical difficulty of the problem, and they were nothing more than a mathematical approach. Hence, there was no idea of representing the universe as a whole by such a simple model as a space filled with a perfect, homogeneous and isotropic fluid. This was just a working hypothesis; and, of course, the more solutions were found, the more readily were the over-simplifying hypotheses discarded. This is to say, attempts were made to construct ever more realistic universes, more in conformity with the observed data, and less schematic. It is thus obvious that if these models were derived from cosmological principles, it was only as working hypotheses. As soon as they were able to, physicists adopted slightly less simple cosmological hypotheses and principles, discarded the continuous, homogeneous, isotropic fluid and tried to introduce rather more complex motions than simple expansion or rotation. And this process is still going on today.

One idea I would like to emphasize is that the point of view adopted by today's cosmologist is that of an astrophysicist wedded to excperience; by which I mean that he eschews all metaphysics and is careful not to extrapolate his fragmentary knowledge of an accessible sample so as to embrace the universe as a whole (not merely a spatial but also a temporal whole). In other words, he shies away from any nonsense about time past and time future in the universe, simply because he lacks the knowledge to make this extrapolation. The mere fact that he now knows how a small sample of the universe behaves does not qualify him to deduce what happened in the dim and distant past, nor equally, what will happen in the future. There is, indeed, far too much taken for granted in extrapolations to the past and future of this universe. To take up Mr. Poirier's example, just as it is difficult to reconstruct a photograph of a human body from an X-ray, so it is difficult — very difficult and even unscientific — to deduce the evolution of the universe as a whole from the scanty knowledge we now have.

I also wish to make a point concerning Gödel's paradoxical universe, much prized by philosophers because of the paradox it contains. Gödel's universe is a model constructed by a highly intelligent logician, well informed about relativistic problems. His model was a description, based on Einstein's theory, of a universe moved by a simple rotational motion and, considering this, he came up with the consequence that in such a universe there are both closed loops and open-ended lines of the order of time, but that to travel along these lines of the order of time is virtually impossible.

But because such a journey is at any rate logically possible, there is a difficulty which Einstein himself discussed. Einstein had taken an interest in this paradox and hoped to see it eliminated by reasons of a physical kind. Work has recently been resumed on this question, that is, attempts are being made, always within the limitations of a realistic cosmology, to see what would happen if part of the universe were to consist of zones in rotation and part of zones in expansion. Obviously this sets some rather hard problems from the topological point of view, and these are being examined and may even be on the way to solution. All that can be said is that taking a much more realistic frame eliminates the paradoxical difficulties encountered by Gödel's model; and so a difficulty of principle which people thought they had to face turns out to be non-existent in nature.

My last point is that this is how modern cosmology sees itself at present: it is not the task of the theoretician to choose for reasons of principle between the various cosmological models put forward, but the task of the experimenters to select among the various possibilities the one that fits the facts best. Today certain experiments have enabled us to extend the area of the universe accessible to our observation; in other words, we now have at our disposal a much larger scale, a much larger explored area, and hence we have experimental data which will perhaps help us to select, or at least, as Mr. Merleau-Ponty said, to eliminate certain models.

At the present time the cosmologist is moving closer to the astrophysicist and further away from the philosopher and theoretical mathematician; I could even say he is becoming something of an engineer. Nowadays many cosmological calculations are carried out like any technical calculation: first they try to account for the experimental data and to interpret them in accordance with the theory of relativity on the one hand, and the other theories resting upon thermodynamics or the evolution of matter on the other. And so it seems that the narrowing of the gap between the theoretical cosmologist and the engineer will open up a new approach — at least, this is how it looks to us at present—and it is along this approach that the metaphysics should be constructed.

V. Kourganoff I must confess that Miss Mavridès has allayed my fears as, listening to Mr. Poirier, I was beginning to wonder whether being a physicist I was still human and my activities still "humanistic". Then again, listening to Mr. Lichnerowicz, I was wondering whether being an astronomer I was a poet or a mathematician, which though attractive would be somewhat inaccurate.

First, I shall take up Mr. Merleau-Ponty's statement: he said, speaking of Gödel, I believe, "Time is not real". But for me, as an astrophysicist, time is not just a philosophical concept, but something physical and concrete. Thus, among several possibilities offered by physics and astronomy, there is that of defining or measuring time by the evolution of the galaxies.

There are galaxies which are obviously "young" and others which are "old". The most distant of those we can see are, because of their great remoteness, which is of the order of several thousand-million light-years, several thousand-million light-years old. This provides us with a comfortably reliable, almost "material" time-scale.

To lend point to my ideas, I shall try to report to you briefly some recent results of "cosmological observations" in the sense Miss Mavridès spoke of, that is, observations which we hope will enable us to test and reallocate the various cosmological models.

We have only been able to speak meaningfully of "cosmological observations" for the last four or five years. For the first time in the history of science we can observe "astronomical objects" which are distant enough to allow us to begin to discriminate between the many model universes which have been built up on the relativistic theories of gravitation and which, until such observations became possible, remained mere exercises in mathematical physics.

First of all we have the observations of "astronomical objects", called by the outlandish name of quasars or, even more cryptically, Q.S.S., both being rather unsatisfactory abbreviations of "quasi-stellar radio sources". These objects set innumerable problems, hardly one of which has been solved. In general, they are assumed to be stars, intrinsically very bright, and this enables them to be observed even though they are several thousand-million light-years away, at the farthest confines of the universe which can be reached by our instruments.

Quasars appear on the photographs taken by optical telescopes as "points", in other words, as stars. However, their spectral lines are even further displaced towards the red than those of distant galaxies. Now, we know that such a shift is interpreted as meaning that the galaxies are moving away from us (Doppler-Fizeau effect). This movement obeys the law discovered by Hubble: the speed of light increases proportionally with distance at a rate of 23 km/s per million light-years.

Three out of four specialists consider that the red shift in the spectral lines of quasars is "cosmological", that is, it means that quasars are taking part in the "expansion of the universe" which Hubble's law describes. According to this hypothesis one can deduce how far away the quasars are from the degree of red shift in their spectral lines by applying Hubble's law in the appropriate extrapolation. This method of estimating their distance, which situates the quasars at distances of the order of five thousand-million light-years away, leads us to attribute to them on account of their brightness (energy received per square metre per second) an optical "power" one hundred times greater than that of the brightest galaxies (although these are made up of billions of stars): it is like comparing a 1000-watt lamp with a 10-watt lamp.

Since May 1966 we know of other astronomic objects in addition to Q.S.S.s. probably equally distant and equally powerful, but which are distinguished from quasars by the distance (or at any rate the weakness) of their radio signals; these are characteristic of Q.S.S.s in addition to the optical radiation and facilitated their discovery. The new objects are called Q.S.G.s and they seem to be about 500 times as numerous as Q.S.S.s.

If the red shifts of Q.S.S.s and Q.S.G.s are truly "cosmological", the study of their distribution in the universe and of these movements should, once we have discovered enough of them, make a decisive contribution to the testing of the various relativistic models of the universe; since the objects (Q.S.S.s and Q.S.G.s) so far observed appear to have an age which equals 0.93 of the age of the universe. (By age of the universe we mean, in this case, the time which has elapsed since what the Americans call the "big bang" — the start of the expansion described by Hubble's law. This does not, however, exclude the possibility that there never was any "big bang". Supposing the speed of expansion to be limited by the speed of light, which is 300,000 km/s, we deduce from the value quoted above for the correspondence between the increase in speed and the increase in distance, an age of thirteen thousand-million years.)

As a by-product of the intensive study of the physical properties of quasars, we may hope to gain an idea of the state of the universe, its matter and its radiation, when it was "young", at the very remote era when the quasars emitted the light and radio signals which are reaching us today with a "lag" of several thousand-million years.

Physically quasars behave neither like stars nor like normal galaxies; they appear to be the site of extremely violent happenings, probably a hundred times more violent than the most powerful atomic explosions.

In the nuclear processes known to us at present (A and H-bombs, stellar radiation, nuclear reactions) only 1% of the mass of reacting particles is converted into energy. In the quasars we appear to be confronted with the complete annihilation of the mass of particles. The processes involved are certainly very different from those which explain stellar radiation.

As the brightness of the quasars is not constant but varies rapidly and widely, it is not yet absolutely certain whether the huge "cosmological" distances attributed to them are genuine; indeed, despite their punctate appearance, they might possibly have a large enough angular diameter for their dimensions to be, at a very great distance, too large for such rapid variations.

Fortunately, in view of the uncertainty about the interpretation of their red shift, quasars are not the only elements of cosmological information supplied by modern astrophysics.

In fact, a quantitative analysis of the stellar spectra enables us to determine the proportion of the two main "chemical" constituents of stars, which are hydrogen and helium. This proportion is found to vary from one star to another, particularly with the age of the star in question, but is always between 8 and 18%: helium is always at least 8% of the stellar mixture. But it has been shown that the nuclear "combustion" of hydrogen in stars should not produce on the surface (which is the only part accessible to direct observation) more than 1% of hydrogen at the maximum.

It has therefore been asked if this abnormally great abundance of helium relative to hydrogen could not derive from an "initial state" of matter, highly condensed and very hot (a temperature of the order of ten thousand-million degrees), a state in which hydrogen might have been converted into helium before the appearance of the universe as we know it today.

Unfortunately this hypothesis, too, is not without its problems, a relatively simple calculation, indeed, showing that if it were adopted the "initial abundance" of helium would have to be 14% *or more*. Now, as I mentioned a short while ago, there are stars which have an abundance of helium *less than* 14% and going down to 8%. The study of cosmic rays gives somewhat similar results, with an abundance of helium going down to 9%.

So, even if we are not yet in a position to draw definite conclusions from this sort of research, it is likely that investigations along these lines will soon enable us to learn whether the universe passed some thousand-millions of years ago through a very dense, very hot stage.

However, as of now, research on the formation of chemical elements from the "primeval mixture" of elementary particles (protons, electrons and neutrons), as well as research in progress on a sort of cosmic "background noise" observed by radio-astronomers, seems to be moving towards the conclusion that some ten thousand-million years ago the universe was about a thousand-million thousand million (10^{17}) times *denser* than it is today.

This is a finding which, despite all the reservations imposed, is something more than mere philosophy or pure poetry.

Of course, the idea that the universe passed through a stage of extreme density does not necessarily imply the "big bang" or a "primeval atom". This stage may well have been preceded by numerous oscillations, and there is nothing to stop this happening again in the distant future.

All we can say is that, by going back thirteen thousand-million years, we do not find the "creation" but the "recasting" of the world, its return to the melting pot, and a revival of the expansion we see today, together with the formation of such structures as stars and galaxies.

Before I sit down, there are two comments I wish to make. I disagree completely with what I took to be the central theme of Mr. Poirier's introductory talk. Listening to him, one might suppose philosophy to be a necessary preparation, a sort of foundation for valid scientific research. I really do not think a man of science can accept such an inflated claim on behalf of philosophy.

It is not the job of philosophy to establish the basis for science. Science and a healthy philosophy should share a common basis: the confrontation of our concepts with experience, with due allowance for the way our minds are moulded by certain failures of experience to conform to our notions, to certain "accepted ideas".

We cannot speak here of "gratuitous speculations". According to Mr. Poirier, we ought to construct mathematical systems which correspond "in little bits" with reality. I say, "No!" When Newton advanced his law of universal gravitation, there was no question of its conforming "in little bits". Newton's law is verified in the solar system, in twin stars and multiple galaxies. As a consequence, the great scientific theories have a purely scientific basis and do not just touch reality here and there; they allow us to make a grand synthesis, if not of the whole of the universe, still of some sizeable "bits".

Moreover, though for a philosopher like Mr. Poirier, or a mathematician like Mr. Lichnerowicz, the universe may consist of "material points", which are the galaxies, for an astronomer galaxies are not points, neither geometrical nor material points. Galaxies and the varied "astronomical objects" they contain are for the astrophysicist highly complex physical laboratories.

The real problem for the astronomer is neither metaphysical nor purely mathematical: it is to incorporate as many astronomical facts as possible into "terrestrial physics" and vice versa. He starts out from the physical laws discovered by the physicists in their laboratories and he examines them to see how they can help him to understand the astronomical phenomena.

The classic problem of the movements of the planets which had dominated nineteenth century astronomy was displaced around 1930 by that of the structure of the stars, and about 1940 by that of nuclear reactions in the heavenly bodies, with the glimpses this afforded of the problem of the evolution of the stars.

The cosmological problem, seen as a problem of the movements of punctate galaxies (1920 model of the universe), is about to be superseded by that of the physical state of the universe, past and future. As I have just explained, what is preoccupying the astrophysicists at present is in particular the problem of the "pre-stellar" state of the world. And in this field we are not making fictitious experiments but real observations, and these observations in increasing measure make it possible for us to choose between viable and non-viable scientific theories.

F. le Lionnais I take the liberty of noting that in my opinion when the scientist is doing his job properly he needs must have something of the philosopher about him which more or less forces him to go in a particular direction. After that, when science has made progress, philosophy must, of course, take note of the fact. But there are really two meanings to the word philosophy: the deep psychological attitude of the scholar and the results of science.

Andrzej Trautman Nowadays we are accustomed to talk of Einsteinian cosmologies, that is to say, those which derive from Einstein's theory of general relativity. Even steady-state theories, and even those of Jordan, are founded upon Einstein's theory.

However, I think it should be mentioned that there is also a Newtonian cosmology, founded upon Newton's theory, with its absolute time and its geometrical structure, which is in effect a natural generalization of the law of universal attraction. This cosmology has very simple logical bases and gives the same main result as the other one as far as the evolution of the universe is concerned. It also leads us to assert that the galaxies are rushing away from us, and the law of distance is the same as in Einstein's theory. I am stressing this point, not because I do not believe in the theory of general relativity, or because I consider the Newtonian theory superior, but in order to convince those who do not support Einstein's theory that the fundamental result of the flight of the galaxies is independent of the individual features of the cosmological theories which support it. It can be obtained by the Newtonian theory. Furthermore, an analysis of Newtonian cosmology gives various results which are interesting from another point of view; it enlightens us about the problem of the ether in the dynamic pre-relativistic theory, defining it as a directional field normal to the hypersurfaces of absolute time in the Newtonian theory. And this does not mean that I subscribe to the ether theory.

This analysis of the Newtonian theory raises a rather more general problem, linked I believe with the general subject-matter of this debate — the synthesis of the exact sciences. There is a program which is currently fashionable in mathematics, of analyzing the fundamental structure of mathematical theories; this program is particularly followed by Bourbaki's group. It involves seeking the basic structures in each mathematical theory, studying them individually and then studying the relationships between them. I think that a similar study might be applied to the fundamental structures of physical theories. In mathematics the word "structure" has a well-understood meaning. In physics I use it rather more loosely, but I mean that we should study the relationship of each element of the structure to experience. The most important revolutions in physics have involved either determining the physical significance of certain structural elements more exactly, or generalizing them, or even changing them. For instance, moving from Newton's theory to special relativity really consists in denying the physical significance of absolute time.

I should also like to say something about Gödel's model. I think mathematicians are careful to exclude absurdities. In Gödel's model it is impossible to formulate a global Cauchy problem because this model of the universe contains no sections of global space. It is impossible to carry out the fictitious experiment of the astronauts circling through the past; there being no longer any notion of state, no initial state can be formulated.

Werner Heisenberg In order to open the discussion on this subject, I would like to start not with opinions or statements, but with questions. Let us now assume that we will try to write on paper some law which, like the Schrödinger equations of old times, will explain the spectrum of the elementary particles. What will we have to know to do this? I think there are essentially two questions which we have to answer.

One question concerns the symmetry properties in nature. The physicists learned from the mathematicians about forty years ago that a law of conservation is a symmetry property, a group property of the underlying natural law. Therefore we have to look for the symmetry properties, and if we know the complete group structure of the underlying natural law then I would say that we almost know the law itself. There may be a few other points to it, like relativistic causality and so on, but that is perhaps not so difficult. That is problem number one, group theory, and I will afterwards say what my own answers to this problem are and I hope that in the discussion there will be, perhaps, very critical remarks about just this statement.

Question number two concerns the mathematical tools which we have to use. Einstein tried to use a field equation. Now, in our time it would certainly be a quantum field equation, namely a field equation connected with all the problems of quantum theory, commutation relation, indeterminism and so on; but that is open to doubt — that is, it may be that the only mathematical tool which can be applied is the so-called S-matrix, or scattering matrix, and the assumption that there exists a scattering matrix is somewhat less narrow than the assumption that there exists a field. So we can either believe that we can work only with this general S-matrix, or we can be more courageous and say — well, we can use a field, or we can pin the thing down on a still more narrow space by saying it must be a field operator in a Hilbert space with positive metric and a Hilbert space which can be constructed from the asymptotic operators alone. The last assumption is of course the one which is closest to earlier quantum theory, and I think it is the one which has been proposed by Wightman, Lehmann and others.

So these are the two essential problems, and it is my conviction that when one has given the answer to these problems one has actually formulated the unified field theory — or unified theory, possibly, but not field theory.

Well now, concerning these two problems I will also briefly state my own answers and, as I said, I very much hope that there will be discussions about it.

Problem number one was group theory. Now, for the group theory the trouble comes in the following way: if there were only exact symmetries in nature, then it would be simple — we would, from the selection rules in elementary particle physics, just see what the exact symmetries are and then we could formulate these symmetries. Unfortunately, there are approximate symmetries in nature, and where this is so, we have to choose in every case between two possibilities. Possibility number one is that the symmetry is an exact symmetry which, however, later is broken by an asymmetry of the whole universe — of the "ground-state" universe. Possibility number two is that it has been an approximate symmetry from the beginning — that is, it is not a real symmetry; it comes in as a rather rough approximate symmetry afterwards by means of dynamics. Well, there is fortunately an experimental criterion which distinguishes between them. In possibility number one — namely, exact symmetry destroyed by the universe — there must be, according to a theorem of Goldstone, bosons — Bose particles — having rest mass zero. In the other case such particles need not exist. Therefore, our own answer is that the Lorentz group, the isospin group and the few gauge groups are the essential continuous groups, but that the groups SU_3, SU_6 and SU_{12} are not fundamental groups in this underlying law.

Then, question number two. We believe that one can formulate a field operator, but that one cannot do it in a Hilbert space with positive metric, that one must allow a wider Hilbert space with indefinite metric, as in the Bleuler-Gupta version of quantum electrodynamics. Again, this answer is open to discussion and one may have arguments for other views. But I would like those who speak in the discussion to try to formulate their own answers to these two typical questions, because I feel that, when one has the answers, then one has almost already the complete unified field theory. That may perhaps serve as an introduction.

Bernard d'Espagnat Professor Heisenberg's opinion about the need to find sooner or later an explanation for the spectrum of particles is certainly well founded; I think we are in this respect rather in the position we were in before the advent of quantum mechanics, and even before the advent of the old quantum theory. The physicists of that time had quantities of spectral lines and could not account for them. I think — and I believe we all feel this way — that it is a major problem and this is one reason why most physicists do not attack it head on. They try to tack round it, to fiddle things a bit; in other words the theories they construct are mainly phenomenological and, one has to admit, nearer Rydberg's work than Bohr's.

As regards the specific questions asked by Professor Heisenberg, it is difficult to make up one's mind. Between the tendency to describe phenomena by the field theory, as was done ten or fifteen years ago, and to explain them by the S-matrix, which has now been in vogue for about five years, we can see a sort of intermediate position emerging, which consists of describing the reactions — yes, certainly — but not in terms of fundamental fields, with the exception of the electromagnetic field. The other fields are not written explicitly; what are described are currents. Therefore the expression of the currents as a function of the fundamental fields is not given. It is possible to introduce the various postulates of symmetry as postulates bearing essentially on these currents, and by writing the problem in this way, we may be able to explain a certain number of things, limited perhaps but by no means negligible, such as the non-renormalization of the vector current, or the calculation of the renormalization of the axial current, or again, the relations between the disintegration rates of various particles and so on. To sum up, there are a number of results which we cannot at present see our way to explain merely by a formalization of the S-matrix. On the other hand, we can obtain them by a formalism to some extent intermediate between the pure S-matrix and the explicit description in terms of fundamental fields. There will still be a field theory but we shall not be faced from the outset with the choice of which fields to treat as elementary, and I consider this a great advantage.

W. Heisenberg Could I perhaps just ask Mr. d'Espagnat a few other questions? If I have understood your French correctly, you suggest a kind of intermediate mathematical scheme, using the currents instead of the fields, but do I understand correctly that you mean that these currents are localized operators acting on something like a Hilbert space, that you actually do introduce a kind of Hilbert space on which these operators, called currents, would act?

B. d'Espagnat More or less, yes, although I do not advocate it as a really fundamental theory. I think that this is still in the realm of these phenomenological attempts that everybody is trying to make. I do not think that kind of thing can bring the final answer, but still, it is a point of view which helps in effectively arriving at some results, which I think pure axiomatic theory does not. I was thinking of this particular thing that I mentioned. But the details — I don't know how I would incorporate such a thing in a full-fledged fundamental theory.

W. Heisenberg My impression would be that the difference (between field theory and current algebra) is very small, unless you say that your current operators are constructed from the asymptotic fields but not from the real fields; but — so far as I understood — that part of this current idea which goes beyond asymptotic theory is just the part which cannot be interpreted as an action only from the asymptotic field, so you actually have slightly more than asymptotic fields. Am I not correct in saying that? So then I would feel that there is not very much difference between these current operators and the complete field operators; but, of course, you may even call the currents a field ... I do not see much difference.

B. d'Espagnat But the advantage is, maybe, that you do not have to specify which fields are fundamental and which are not.

W. Heisenberg That is perfectly right. I would not like this distinction at all between fundamental and non-fundamental fields — it is very arbitrary.

J.-P. Vigier Concerning the nature of present-day problems, I am basically in agreement with Professor Heisenberg. The fundamental problem today is the problem of symmetries. Indeed, the great advances which have taken place in quantum mechanics over quite a long period were based upon a special dynamic symmetry, the Poincaré group, which governs the dynamics of particles regarded as points. Dirac's fundamental contribution was quite simply the linearization of the first invariant, of the first Casimir operator of the Poincaré group, the square of the mass.

We thus have a situation of the same type as the Balmer or Rydberg series, where we observe a whole series of mass levels in elementary particles; the question is to know what lies behind this appearance and, above all, whether there is a group of fundamental symmetries.

My reply to this question would be different from Professor Heisenberg's. I would say that I do not think that the symmetry of the particles of the SU_3 type (or whatever symmetry group finally triumphs) can be dissociated from the Poincaré group. What we really have is, on the one hand, a group which governs the external movement of particles — the Poincaré group which contains mass and spin — and on the other hand a series of new quantum numbers, isotopic spin, hypercharge, etc. Thus the question is to find out whether all these new quantum numbers are part of a deeper global symmetry. My answer would be that, if the description of physics we have accepted until now is true, there must be a unified group underlying this appearance.

There have been some early attempts at extension, for instance, Michel's; he tried to extend the Poincaré group by a number of gauge groups, but, in fact, his mathematical procedure produced the extension only via what are called Abelian gauge groups.

Quite recently an outstanding mathematician at the Henri Poincaré Institute, Flato, put forward another idea which involves combining the Poincaré group and the group of internal symmetries into a global group in such a way as to obtain an underlying dynamics which can explain the behavior of the particles. And indeed, in this case the mathematical result obtained corroborates Professor Heisenberg's second hypothesis, the idea that non-compact symmetry groups of indefinite metric are required. Again, the proposed group had already been studied by Einstein as the group of conformable invariances, that is to say systems which are not only at constant relative speed but also at constant relative acceleration; in cases like this we find a symmetry which resembles that of SU_3 but is not identical and which is called SU (2, 1).

I think in the end this question of whether a symmetry exists or not will settle the matter regarding the possibility of a unified theory. If there is no symmetry, there can be no unified theory because, as Professor Heisenberg has said, all the laws of conservation in fact derive from properties, invariances and deeper symmetries.

I think, too, that current algebra is only one stage on the road to the search for deeper symmetries; if certain quantum numbers are conserved in current interactions, then we have in the end a symmetry in current algebra. So this is the problem: is there in nature a deeper symmetry behind the appearance we observe? We have to find the answer to this question and I think that, consciously or unconsciously, all the attempts being made today are really attempts to answer this question.

I shall conclude by quoting a remark of Feynman's which I think very profound: What is the use of introducing symmetries if we are going to destroy them right away? We bring in a fundamental instrument to govern the dynamics and then destroy it as soon as we run up against difficulties.

This is not a coherent theoretical procedure. To put it differently, all the empirical work of unravelling being done by today's physicists is quite fundamental, yet we are still at the stage of the workers who examined the spectrum of the hydrogen atom, the helium atom and so on, and yet were really looking for a more profound dynamics. The one we are looking for today should play in relation to elementary particles the role that Bohr's theory played relative to the atomic spectrum. It is in this sense we must understand the hope of constructing a unified theory of elementary particles.

J. Ullmo I would just like to emphasize that we are witnessing something pretty extraordinary — I think physics is just about to effect a breakthrough towards something quite fundamental, like that Professor Heisenberg achieved some forty years ago and which he may be on the point of doing a second time.

I think his definition of the alternatives with which we are faced is excellent. Fundamental particles have been found to fall into certain groups, or certain symmetries, but not in a perfect manner. There are thus two possibilities: Professor Heisenberg favors one and Professor Vigier the other.

The fact that the world does not conform to the symmetries of the groups at present discovered seems to Professor Heisenberg to be a contingent fact of the initial conditions, of the fact that there is a certain dissymmetry in the world, and that this prevents the underlying symmetries which operate on the objects from being fully manifest. This may look like an introduction to contingency, which is what Professor Vigier said. Why think up this origin of things to break the symmetries which we are going to use later on? But Professor Heisenberg's very clever answer is that we can make use of this contingency. The fact that the symmetries are not respected in the initial conditions introduced those long-range fields which are familiar to us. The electromagnetic field is of this type and I understand the gravitational field is too. After all, the theory of physics rests upon this apparent incoherence and initial contingency and draws some remarkable conclusions from them.

On the other side there are the views which Professor Vigier developed quoting Flato, and these also seem to me excellent. Here, there are no concessions: the fundamental group is respected completely and the apparent breaks in the groups relative to the internal motion of the particles, transforming them from one to another, are simply due to the fact that the general motion in the universe, which comprises simultaneously space-time and the internal space of the particles, is not part of the sub-group of the internal motions and hence does not conform to the invariances of this internal motion. Here, too, the break is put to excellent use — unfortunately it is a little hard to explain — since we obtain relations of mass between all the elementary particles and these, so far as I know, are all admirably verified.

W. Heisenberg I would first like to make a few remarks concerning what you, Professor Vigier, called global symmetry. I agree completely with you that we should start — or that any unified field theory which can be correct — should start with a statement about the global symmetry containing the Poincaré group and other groups. Now, I think the simplest way — or almost the only way which I know — to state such a global symmetry is just to show an object which has this symmetry; and therefore one may also say that the unified field theory starts with something — namely, an equation — and then one has simply to see under what operations is this equation invariant, and that, then, is the statement of the global symmetry.

But, I would like also to say that one should not connect this question with any kind of mysticism or say that it is very profound and so on — it is just very trivial and normal physics. And there we have to ask what are the exact symmetries? That, I think, is really the first question and — now I have expressed a definite opinion — well, that equation which I like to use as a basis for the field theory is exactly symmetrical under the Poincaré group, under the isospin group, under a few gauge groups and under a few discrete groups which we need not discuss now in every detail.

Now, one can either believe that this is already the correct global group or one can believe it is not, and I would therefore very much like to have your opinion. Do you think that these are the only exact symmetries of the underlying law, or do you feel that one should include other symmetries like SU_3, SU_6, and so on?

J.-P. Vigier Just one precise question. There is no great difference between writing down a wave equation and saying it is invariant under a global symmetry group, or starting directly to discuss a global symmetry group, except that this is trivial physics. The point is this: I think the internal symmetry group should also govern the wave equation; in other words, all the quantum numbers should come from the invariant properties of the wave equation; I mean not only the Poincaré symmetry but also isobaric spin, hypercharge, and so on, and even SU_3, if you like. But, as I said, for other mathematical reasons we think you should take a non-compact group with an indefinite metric to classify internal states. Whatever you choose, I feel very strongly that the wave equation should contain everything, because if it does not, then you are not writing a truly unified theory — you have to add something else to the group. And the question of the initial conditions is quite a different question from this one — meaning that, for example, the gauge groups give you the conservation of quantum numbers. However, if you want to conserve isobaric spin, then your wave equation must be invariant under isobaric spin transformations. If you want to conserve hypercharge, the wave equation must be invariant under an Abelian gauge group, and so on. And the whole set of levels — if this is the point of view you have developed, and I agree with the basic idea — should result directly, without any addition, from the fundamental symmetry itself. Therefore, the basic problem facing the physicists now — I do not say whether the answer is correct or not, it is a question to discuss — the basic question is: how are you going to unify Poincaré with internal symmetries? Meaning, if you want to take the word of Michel, how are you going to extend the Poincaré group to include the conservation of the newly observed quantum numbers?

W. Heisenberg My answer would be that I include all those groups, and only those groups which are the groups of that equation which we use; in other words, that equation which we use does contain more than the Lorentz group, it does contain the isospin group, and also gauge groups. I do not think that one should include more than that — such as those very vague, approximate symmetries like SU_3 — because they are not fundamental symmetries. And so I would say: alright, we know the exact symmetries — that is, the Lorentz group and U_2 — namely isospin, gauge and so on; but that is all, and they are all included in this equation. So, by means of this equation, we have connected the Lorentz group already with these other groups. And that, I think, is sufficient — or do you not think so?

J.-P. Vigier No, I do not think it's sufficient, because the notion of approximate symmetry, in itself, has to be clarified. (H. Yes!). What is an approximate symmetry? From the point of view of observed facts, what we observe is the conservation of certain quantum numbers (isobaric spin and so on). Something which is conserved like that should, I think, be tied with the symmetry, the basic symmetry of nature itself. (H. Oh yes!). If you do not do that, then you are in a very difficult position to write an underlying dynamic.

W. Heisenberg But I do it! I only want to say that SU_3 are not as well-defined quantum numbers. If there were a conservation law for SU_3, then you would be right; since there is none — or only a very vague, approximate one — I do not see that we need to express it in the fundamentals.

J.-P. Vigier Then let us drop the argument on SU_3, because I think we agree. In fact, Biedenham has recently shown that you could have similar results, for example, with the group we are using — namely, SU_2. What you need is isobaric spin conservation, and hypercharge, and baryon number conservation, and so on, meaning you can have wider groups or different groups which would give you the similar conservation of quantum numbers; and then, of course, there remains the classification of multiplets. But the essential point is that of unification, because if you do not do a unification, there is no hope of explaining the mass.

W. Heisenberg Yes, but we do unification. I mean, in what sense would we not try unification? That I do not understand.

J.-P. Vigier Usually people now take the direct product of the Poincaré group with some internal symmetry. But if you have a direct product, then there is no hope of splitting the mass of the multiplets because the square of the mass commutes with all the internal symmetry. Therefore the multiplets are necessarily degenerate.

W. Heisenberg Well, of course, this breaking of the groups — for instance, of the isospin groups — that I would not connect with such things, but I would like to connect it with an asymmetry of the ground-state world, with the Goldstone theorem. If you would not allow that, then, of course, you may be right, but I cannot see why you object to the idea that the ground state is asymmetrical.

J.-P. Vigier It is a question of results!

W. Heisenberg I agree, but I think this theorem of Goldstone was an enormous help. It said that if we have an asymmetry due to the asymmetry of the ground state; if originally there was a perfect symmetry, which is later on destroyed by the ground state; then and only then, we observe bosons of rest mass zero. And now we see the bosons of rest mass zero, we see the photons, so I think we have a very good argument in our hands for saying that this is a case where the ground state is asymmetrical.

J.-P. Vigier Yes, but the argument cannot be put in such a strong, sharp form; I mean that the existence of zero mass particles does not prove the splitting of the ground state; you could get mass zero particles from an unbroken symmetry. You see, I think Feynman's argument, from that point of view, is very deep. What you want is a total dynamic, and a total dynamic means a global symmetry group.

Determinism and Indeterminism

F. le Lionnais I would like to call upon Professor Heisenberg to open the debate on determinism and indeterminism by explaining to us his present point of view on this question.

W. Heisenberg I was not prepared for opening the discussion on this rather old problem, but of course we can discuss it in this connection because Einstein's difficulties in accepting the quantum point of view came from his hesitation to accept fundamental indeterminism. So we can, of course, come back to this old question now. But on the other hand, there has been so much said about the question, that I really do not know from what side I should start it.

You know that in quantum theory there has been one interpretation of the quantum theoretical laws which is sometimes called the Copenhagen interpretation, and this interpretation is meant in the following sense: that those mathematical objects which we write down on paper do not give a statement about the objective situation in nature but, on the contrary, they make a statement about potentialities, about probabilities. When we write down our Hilbert state vector in such and such a form, then we say: if something is now done, then this will probably happen, or there is this probability, and so on. So the central point in this indeterministic interpretation of quantum theory is the introduction of the concept of potentiality — as in the old Aristotelian philosophy — or of probability, to put it in more modern language. And the point is that our mathematical scheme refers to potentialities and not to realities; it is not realistic in the old sense. Einstein was not willing to accept this, because he said there must be an objective nature which is there without our speaking about it, without our knowing it; it is there and we must be able to describe it.

That is the problem, and I have stated my own answer; perhaps that is enough to get others into the discussion and to raise questions.

O. Costa de Beauregard I belong to the probabilist persuasion. First of all, I think, in company with Popper and Landé, that classical statistical mechanics, although constructed with a hidden determinism, really inclines towards a theory without a hidden determinism. In order to convey what I wish to say, I shall make use of a parable: as everyone knows, the theory of the Doppler effect and aberration can be constructed with pre-relativistic kinematics, using an ether. The important thing is that absolute speeds are eliminated from the result and only the relative observer-source speeds remain. Hence, although the theory of the Doppler effect and aberration is possible in classical kinematics, the result tends to favor the absence of ether and consequently of classical kinematics. Like Popper and Landé, I think this is a situation analogous to classical statistical mechanics.

I would like to expose the paradox which consists in deducing objectivity in the increase of entropy merely from an ignorance of the micro-conditions. How can ignorance explain perfectly objective phenomena? In fact, what is important is not that we do not know the micro-conditions but what we do know about their statistical properties as a whole. The strength of classical statistical mechanics is that what is hidden is ignored while the emphasis falls upon what is objectively known, that is, the statistical distribution.

In this connection, I should like to recall all the arguments advanced, for example, to establish the Maxwellian distribution of speeds. They are all related in that the hypothesis of a complex statistical chaos is accepted in one form or another. No more than this, but it is enough to give Maxwell's law of distribution. I should also like to recall here Landé's argument about balls dropped on a razor's edge. The balls fall to the left or right in accordance with a statistical law. Landé here makes very effective use of Leibniz' argument of continuity. I experienced quite a shock on reading this, as I remembered that Poincaré had drawn the opposite conclusion from a similar example; this is the idea of a neddle stood upon its point and falling one way or the other. How is it then that from the same material argument Poincaré draws an anti-Leibnizian conclusion of discontinuity between cause and effect, while Landé opts for cause-effect continuity in Leibniz' sense?

I examined both arguments carefully. Of course, neither Poincaré nor Landé had made a mistake in the reasoning, but they started out from different premises: Poincaré postulates a hidden determinism, while Landé assumes a basic probabilism. Please note, I am not arguing against the existence of a classical sub-statistical world, any more than against the existence of a sub-quantic world. What I am trying to say is that I think we should seriously question the need to suppose that this world is governed by deterministic laws.

And now, if we move on to quantum statistics, what is essentially new in it? It is certainly a fact that the pure case shows dispersion relative to certain measurements we can make. And it is certainly to this that we owe all the other fundamental properties of this new statistics, such as, for example, the interference of probabilities, which has been most lucidly analysed in Landé's new book.

I should also like to say that I am far from being in complete agreement with all that Landé says. His little book contains some things one could not possibly agree with, yet along with these come other extremely penetrating things.

Another point about quantum theory which is essential to this discussion is the disturbance linked with measurement. I think this is a most valuable feature and, in order to explain it, I shall have recourse to another modern discipline, cybernetics. Cybernetics introduced the concept of information and showed something that had not been suspected before, namely that every time a piece of information is acquired about a physical or even a non-physical problem, whether by computer, telephone or what-have-you, this information always produces a corresponding fall in the negative entropy of the ambiance. But cybernetics has gone further still; it has shown that information is not only an increase in knowledge but also a means of intervention, a means of restoring order in the cosmos in the sense of macroscopic statistics. It has accentuated two symmetrical aspects of information, information as an increase in knowledge and information as a power of organization. Thus, without seeking to do so, it has chanced upon an old Aristotelian association between the two symmetrical aspects of information.

The fact that the quantum theory links with the acquisition of knowledge about a system a disturbance of this system is extremely precious from this point of view, because it indicates an essential bond between these two aspects of information, so that one cannot in the end separate them. I think the discussion, for example, of the process of measurement in quantum mechanics would gain a great deal from the introduction of this word information. It is, of course, implicit throughout von Neumann's book; perhaps it ought to be made more explicit. Seen in this light, the reduction of wave bundles which has caused so much ink to flow is not such a catastrophe; it is the passing of the information which is acquired every time we make a statistical test, it is the passing of the information revealed when we turn over a playing card. I think the waves of quantum mechanics are indeed waves of probability. As Landé says, "they are tables, just like an insurance company's mortality tables".

Before I close, I should like to say another word on a matter which has been discussed at length here: the quarrel between modelism and formalism. It is always said, and with reasons, that the success of the classical statistical mechanics underlying phenomenological thermodynamics is a great triumph of the model approach. I should like to say that one can quote other cases, just as striking, where the verdict goes the other way. There are cases where the formal approach has been more fruitful, the typical case being precisely that of the theory of special relativity where the formal approach triumphed over the model approach of the ether theories. And again, if you open a treatise on classical statistical mechanics, you will be surprised to find how little it uses the model approach, being on the contrary substantially formalist and abstract.

W. Heisenberg I am not sure whether I have understood everything on account of the language, but I have understood — or believe I have understood — that you feel that one should rather interpret quantum theory along the lines of the papers of Landé. He was interested to say that one should compare the laws of quantum theory with the laws in statistical mechanics, correlation, and so on; therefore it would not be very wise to use those general philosophical terms which the Copenhagen interpretation sometimes uses.

I would like to make the following remark. I understood — but perhaps I am wrong here — that this interpretation of Landé, or the interpretation you suggest, will not disagree with the other interpretation concerning any kind of predictions; so, whatever the predictions for a future experiment might be, the two interpretations will always give identical results. Let us assume that this is so — if not, I hope you will interrupt me. Then, of course, it is a question of language, a question of what words should be used to describe this rather strange situation which we apparently know because we can predict everything. The next question is: is this language — which, for instance, Landé uses — a convenient language? According to my view, it is not, for the following reason: the mathematical scheme in quantum theory shows enormous flexibility — that is, the same mathematical scheme can either be called a scheme working with coordinates, momenta of particles; it can also be called a mathematical scheme which just by some transformation goes over into another mathematical scheme speaking about waves and amplitudes and energy densities and so on. So it is quite obvious that this mathematical scheme can be connected with classical interpretations in many ways; it has enormous flexibility. But if one uses a language of the kind which Landé uses, then one loses this flexibility, one is, for instance, forced to say: alright, we have a statistical mechanics of particles — namely, the electrons in an iron atom; they are particles having a specified momentum, having specified co-ordinates; and then we have the statistical mechanical scheme just of this thing. But that is much narrower — much too narrow for the mathematical scheme: the mathematical scheme allows different interpretations. So I would hesitate to apply a language which is too narrow, which has not the same flexibility as the mathematical scheme. Actually, the mathematical scheme describes everything which we need to know, because we can predict any experiment. But then, if we translate this mathematical scheme into ordinary words, then, of course, we have the difficulty that our ordinary words just do not fit in with this mathematical scheme; and so we have to make compromises. But I must say that I dislike a compromise which makes the whole thing very narrow, which just shows the way to some part of the mathematical scheme. And that is the reason why I do not feel happy with these rather narrow interpretations.

A. Matveyev First of all, I should like to repeat the statement I made during my intervention in an earlier debate, as I think it is applicable to the problem under discussion now: namely, that we try to understand what is going on the world with the aid of ideas and notions which exist in our heads; but these ideas are the result of our previous attempts to understand the world, whereas the ideas we have at present are the result of our macroscopic experience and we may be sure that they are applicable only to macroscopic experience and to low velocities.

If we begin to study high velocities and microscopic objects, we are not sure that these notions are applicable at all — we are not sure that they make sense in this new field. This is clear. When we came to high speeds, we were obliged to create quite a new picture of space and time in the theory of relativity. When we came to consider microscopic events, we were obliged to elaborate new approaches. We know that there are neither particles nor waves. There is something synthetic which shows itself as a particle or a wave depending on the conditions of "the show". This is a general approach. Therefore, when we consider the problem of causality, we must first of all analyse what we mean by this word and whether it should have universal meaning, or whether it has any meaning at all in microscopic theory.

What do we mean when we speak about determinism and indeterminism? Very well, we have a concept of determinism which was formulated in our macroscopic experience. What are the essential features of this macroscopic experience from the point of view of determinism? First of all, we suppose that sequences of events in time have a meaning. This is the first assumption. Thus we may say that this is one event, this is the second event, this is a cause, this is a sequence and so on, and everything is alright — we understand what we mean by causality in a macroscopic theory. But are we sure that the same meaning should be attached to it when considering microscopic experience? No, we are not sure because, first of all, the meaning of the description of the state is quite different. Then, we are not sure at present that the notion of sequences of events has meaning, because in statistical theory, we often cannot speak about sequences of events except in asymptotic regions.

In these circumstances, how can we speak about causality in the sense of macroscopic experience? In order to use words which have some meaning, I think the first step is to define what we mean by determinism in our microscopic experiments.

In principle, I agree with the Copenhagen interpretation of quantum mechanics, but I cannot agree that this interpretation means indeterminism. It appears to do so because indeterminism is taken in the old sense of macroscopic experience. Without going into details, we can state that causal relationship always exists when there exist laws which permit us to predict the development of events. It should also be noted that the notion of causal relationship is much wider than the notion of relationships described by physical laws. The aim of scientific research is to discover laws, that's all. The existence of laws which enable us to predict development is the expression of the existence of causality. Of course, the meaning of causality for statistical laws is different from the meaning of causality for dynamic laws. The meaning of causality also depends on the characteristics of description.

Let us take an example. In macroscopic experience, we can identify different parts of a system and can also identify in time the changes which may take place in different parts of the system. It is easily seen that this is a precondition for the possibility of causal description in the macroscopic sense. Such identification is impossible in microscopic experience. Therefore a causal description in the macroscopic sense has no meaning in microscopic experience. Supporters of indeterminism in microscopic experience draw the following conclusion at this moment: the observation of the above-described situation is equivalent to recognition of indeterminism in microscopic experience. But, logically, this is a wrong conclusion. Logically, the right conclusion is the following: acknowledgement of the above-described situation would have been equivalent to recognition of indeterminism in microscopic experience if all the other characteristics of the microscopic experience had been identical to the corresponding characteristics of macroscopic experience. But such is not the case. From this point of view, acknowledgement of the above situation is not the end of the analysis but only the beginning. That is why I cannot agree that the Copenhagen interpretation of quantum mechanics means indeterminism.

W. Heisenberg I would like to make only one modest remark about these terms "determinism" or "indeterminism". It is quite clear that one can use these words in many ways, and therefore they are not very well defined; but just from the modest point of view of a physicist one would say the following.

For instance, in old statistical thermodynamics, we said: "here we have a system but we do not know the coordinates well enough, so we cannot predict what it will do after such and such a time", then we had the idea that *in principle* we could determine it because in principle we would be able to measure all those positions of the particles and then we would be able to predict. Now, in quantum theory we have encountered a new situation. When we have a radium atom which will possibly decay with emission of an alpha ray, then we can again ask: "can we possibly predict when it will be emitted?" And the answer is, we cannot predict; and there is now this difference: even *in principle* we have no hope of making any kind of measurements of this alpha particle which would allow us to predict its decay. So there is a difference between quantum theory and the earlier theory in the sense that we have given up looking for parameters which we could measure, and thereby predict more than we can already by quantum theory. That is the only way in which the quantum physicist usually uses this word indeterminism. Whether that is a good way of using the word, I do not know.

O. Costa de Beauregard I should like to clear up two misunderstandings. First, I do not set out to defend Landé. I said there were some questionable things in his book, but that along with these there are some that are very interesting and make one stop and think, particularly his way of deducing interference between probabilities which is most ingenious.

Then, I certainly was not trying to reduce quantum statistics to classical statistics. I said in so many words that it contained something quite new, the pure case with dispersion; I therefore wholly support formalism and the Copenhagen School, except perhaps as regards the words I use. Basically, as regards the mathematical interpretation, I am in complete agreement with the Copenhagen School. I did not say that quantum statistics goes back to classical statistics, I merely said that classical statistics might incline one to think of the existence of a basic probabilism in just the same way as, once one has done the theory of aberration and the Doppler effect in classical kinematics, one can say that since absolute speed is eliminated, it would be better to have a theory which does not include absolute speed at all. This was all I meant to say.

J. Ullmo I think this debate is rather old hat. We know that classical determinism is a limit case and that we do not see it either in quantum mechanics or in experience. For this reason, quantum mechanics is in perfect conformity with experience and the example of a radioactive atom is absolutely legitimate; but the very clever interpretations of Professor Heisenberg have shown us that the inability to predict comes from the limitation of our knowledge and that this limitation is, morcover, insoluble because of the very conditions of experience. Thus, we must give up determinism in the sense of perfect prediction because this is the only way we can preserve the much more important idea of causality. Not causality in the meaning Bohr gave to it formerly, which was simply the application of rules to classical mechanics, but causality in the philosophical sense, which comes down to a single statement: two identical systems evolve in the same way unless there is a reason why they should diverge; and if there is, then they are not identical. So what are we arguing about? It is not very desirable that we should do away with the principle of sufficient reason, or causality; and it is technically unnecessary. What is more, classical statistical mechanics could just as well be built up on a basis of pure chance—Langevin told me this thirty years ago! But it is extraordinarily difficult to do away with the insistence on the sufficient reason, because the whole of science and the whole of reason are built upon this rock.

Finally, another point which interests me is the opposition between formal and model theory. I would really rather call the latter structural theory, since to talk of models conjures up nineteenth century pictures of Lord Kelvin constructing mechanical models. The real difference is that formal theory has nothing but its success to go on, whereas structural theory believes in the reality of what it describes. This reality may be of an extremely abstract mathematical kind, but it is set up as an objective structure.

We have heard here, in the debate about Einstein's thought, how he started out from a positivist attitude, corresponding perhaps to a formal theory, but was led by reflection to a very profound structural theory; I think the course followed by Professor Heisenberg's development, which is quite as significant as Einstein's, has been much the same.

A. Matveyev I should like to make a short comment on the interesting remark made by Professor Heisenberg. He has explained what we mean, speaking about indeterminism, when considering a problem of defining the exact time-point of radioactive decay of a nucleus. We cannot in principle predict the time of this event, so this means indeterminism. My answer to this is that the question about determination of the exact moment of the transition of the nucleus seems to me to have no physical meaning.

I will explain. In the history of science, questions have sometimes been asked which have no meaning and require no answer. For instance, some two thousand years ago, Aristotle asked what causes particles to move at a uniform velocity. For two thousand years, scientists tried to answer this question. As you know, Aristotle himself introduced a special force which was responsible for the uniform motion of bodies. A lot of papers were written dealing with this problem and trying to clarify its solution. But then along came Newton and said that there was no physical meaning in this question, that there was in fact no question, so no answer was required; that this is a basic law of motion requiring no further explanation.

There are many physical events with no physical law behind them. For instance, a man is struck by a car at some point in a street. Consider this event. We may follow a chain of many other events which resulted in the man's arrival at the point of accident at a particular moment. On the other hand, we may follow a separate chain of events which resulted in the car's arrival at the point of accident at the same moment. It is clear that there is no physical connection between the first and the second chain of events. By absence of physical connection, I mean that changes in the first chain of events would have had no effect on the second. Thus there is no physical law behind this accident. A question about the prediction of the time and place of the accident has no physical meaning. On the other hand, no one would deny the full causality of the sequence of events under consideration.

An analogous situation exists in the case of radioactive decay. At present, at least, we do not know of any physical connection between radioactive decay and other chains of physical events. Therefore the situation in this case is absolutely analogous to the example I have just given. This is why the question about prediction of the time of decay has no physical meaning. On the other hand, the impossibility of predicting the time of decay does not prove indeterninism for this event.

W. Heisenberg In general, I agree completely with every word Mr. Matveyev said, in the sense that we very frequently use words which are not well defined and therefore many misunderstandings arise because we think that we know what we mean by the words and actually we do not. So with this general attitude, I could not agree more. Still, with the problem of time I think it is a little bit more difficult and one must also be careful not to be too ready to reject the meaning of such a statement. What I mean is this let us assume that we have a radium atom here and a counter arrangement, so that, when an alpha particle is emitted, we shall certainly see it in the counter. Then the following uestion has a definite meaning will the counter be triggered before twelve o'clock, or not?

Now, that can happen or it cannot happen, but certainly this question has a very definite meaning. The trouble is that we cannot predict — but the question has a definite meaning. Would you agree about it?

A. Matveyev I could not agree more. It has indeed a definite meaning, but as I have already explained, it seems to me that it has no physical meaning, and proves nothing.

B. d'Espagnat We are in very deep water here! I should like to come back to a rather more elementary formulation of the question and to what Professor Heisenberg said in opening this debate. He spoke of Einstein's idea that there was a reality which existed objectively, independent of the observer, and which would exist just the same if there were no observer. Perhaps we could call this the "realistic" approach. Professor Heisenberg said this was what caused Einsteins's difficulties with orthodox quantum mechanics, as formulated in Copenhagen and laid down in the textbooks.

I think this is the crux of the matter. I believe, in other words, that the question raised by quantum mechanics is not so much one of determinism or indeterminism as of whether our science is compatible with a realistic conception. It should be noted, too, that this sort of attitude is something one can take or leave. Great physicists like Niels Bohr did not have it. In passing, I should like to quote Bohr: "the sole object of quantum mechanics is to describe the observation accurately." Such an attitude carried to its logical conclusion reduces science to the mere prediction of the results of future observations from the results of past ones. Looked at from this angle, quantum mechanics as presently formulated is perfection, beyond reproach; it works very well and is the simplest kind of formalism that can be invented. Thus we may judge that such a pragmatic approach is a very healthy one for a practising physicist.

But we ought perhaps to look at a few philosophical problems, too. In this sphere quantum mechanics looks much more satisfactory. It does not accept the taking into account of separate objects or the knowledge, however approximate, of particular considerations. More especially, it allows the hypothesis, which is after all quite natural and, in my view, justified, that there is a *reality* underlying the phenomena. But after that it becomes very difficult to avoid making, if only by implication, hypothesis about this reality which are unduly naive, by which I mean that certain observable consequences might be able to contradict the predictions made from them.

Thus, on the microscopic scale at any rate, the linearity of the laws of quantum mechanics generally excludes the consideration of separate objects unless they have never interacted in the past. Without going into the rigorous developments which the subject really demands, one may imagine without too much difficulty the immense consequences which this simple fact of inseparability could induce in our vision of the universe. So it happens that, if quantum mechanics still contrives to some extent to preserve our customary notion of the object, it is only at the expense of the "pure, hard" conception of objective reality such as Einstein, among others, possessed and which he always stubbornly defended.

But, as I have already said, so long as we restrict ourselves to a simple description of the sequence of phenomena, questions like this do not arise and the linear laws of quantum mechanics are perfectly satisfactory.

J.-P. Vigier The difficulty is that we have here representatives of several versions of the Copenhagen interpretation. I am greatly suprised to find my position is much nearer Professor Heisenberg's than Professor Matveyev's. The latter seems to belong to a very idealistic wing of the Copenhagen School.

I shall break the problem down into three parts. The first question, which I consider to be beyond dispute, is the question of the completeness of the description given by quantum mechanics. The entire history of physics has shown that things work the other way round. The further we advance, the more we come to realize that we have to introduce new parameters into the description of nature, beginning with isotopic spin and gauge parameters. Therefore, any claims that the story of physics is finished, put out, moreover, in various guises, and not always by the same people either, are simply not worth discussing; the matter has been settled by the advance of physics itself.

Second question: we are all in agreement that quantum mechanics is a valid statistical description; but is it a complete description or is it not? There is no question of going back to Laplace, classical mechanics is dead as far as the description of microphenomena is concerned. There is equally no question of predicting the onward march of the world, like a machine, to the end of time. The further we go down, step by step, into the infinitely small, the more we see properties emerging which are non-classical and of a type quite different from, for example, the fluctuations of the zero point. We all know that the image of the void which comes out of this is extremely complex and quite unlike anything which went before. We are forced to construct a deeper level and to postulate dissymmetries in the void in order to explain things at the level at present observed.

My second point, therefore, is this: the proposed procedure is completely incompatible with certain positions adopted by the Copenhagen School, as expounded here by Professor Matveyev. One has a perfect right to introduce into physics things one cannot directly observe if it helps to explain the things one does observe. Atomic theory had to wait forty years before it was experimentally established, and the great failure of the positivists was indeed to have denied the existence of atoms on the grounds that they could not be seen; history eventually showed how wrong they were.

Now I come to my third question about the Copenhagen school: here I think the various representatives of this school, even those present at this forum, are not basically in agreement. This is the question of the objectivity of things independent of the observers.

I subscribe to what has been said about actuarial tables; clearly, the death of the insured person is in no way influenced by the tables. But there is another side to this question: are the field equations equally valid whether there is an observer or not? Personally, I think along with Einstein, Mr. de Broglie and others that the answer is "Yes". Dirac's equation was true long before any quantum physicist thought it up and wrote it down. Or again, if I walk through that door, I shall still exist on the other side of it.

We should not, however, be credited with the absurd idea that, due to some sort of bias, we wish to reinstitute in the properties of micro-objects ideas which belong to a totally different plane, that of classical mechanics. The problem is rather to know whether and to what extent we should conserve the properties of relativistic space-time at the micro-object level.

W. Heisenberg I would like to make a few remarks concerning this problem of the completeness of a theory. Of course, this term completeness, or the suggestion that a theory is closed, has been misunderstood many times and I would like to say in which sense I think that quantum theory is complete or not complete.

Let us first speak about the old Newtonian mechanics. Is Newtonian mechanics complete, is it a closed scheme or is it not? I would say — and this may seem to you somewhat paradoxical — it is a complete theory, and it is absolutely impossible to improve it, in the following sense. If you can describe parts of nature with those concepts which are applied in Newtonian mechanics — namely, co-ordinates, velocities, masses and so on — then the equations of Newton are exact equations and every attempt to improve these equations is simply nonsense. But of course there are other parts of nature in which these concepts just do not apply — this is already so in relativity, was already so in Maxwell's theory, where we had the concept of a field, and is certainly true in quantum mechanics, and so on. And in the same sense, too, I feel that quantum mechanics is complete. As long as we can speak about nature just in these terms — as stationary states, transition probabilities and so on — then quantum mechanics gives a complete description. But already, say, relativistic quantum theory — elementary particles — has not been contained in the scheme. But it is very interesting to see that all the development which has taken place in the last thirty years — the introduction of isospin, introduction of strangerness and whatever else — all these new things have not at all changed the principal questions of quantum theory; so actually these discussions which we have now about determinism and indeterminism were held in almost the same form thirty years ago, when Einstein and Bohr discussed it at Brussels. There were the same questions and the same answers were given. And that I think shows that within those concepts which are meant by quantum theory, within those concepts actually, the theory is complete. But of course, now we have a new part of physics.

Now let us assume that this unified field theory will be understood within a few years and that we know the fundamental equation which allows us to derive the spectrum of elementary particles, and again the question arises: is this theory complete? Now, I would agree at once with Mr. Vigier that it will not be complete in the universal sense — that, for instance, biology will not be included; and whatever the answer to the elementary particle problems, the problems of biology will not be solved by these answers.

So it is clear that again physics is open at some other points, but within this scheme of concepts alone I think it is closed and is complete, and I would therefore like to introduce the concept of restricted completeness. In that sense I would say that quantum theory is complete!

Reverend D. Dubarle As a lay follower of the stimulating discussion we have just listened to, I should like to add a few words. I think we are all agreed on the operational aspect, but philosophically there is an underlying problem which did not arise at all in the classical theories if a system coordinates correctly, perhaps even perfectly, what we know we are doing, is it just as true that this is exactly what we ought to be thinking? This should provide us with food for thought, and I believe many of the questions which have been asked here arose out of this uncertainty. We want to know how we can coordinate a certain language — which is no longer that of rigorous science and its methodology — with this system. Professor Heisenberg has insisted most usefully on the language. I venture to ask a question on — shall we say? — concepts. Classically, we maintained causality or, better, the principle of sufficient reason by the type of explanation given in classical analytical mechanics. The problem is: is this the only way of doing so? I believe that many of us are sure there must be other systems of establishing a sufficient reason. Such a system is very important and perhaps ought to be discussed further, but there is a growing tendency to think it is not absolutely unconditional and obligatory. We should be looking for others. We shall probably have to do so at the time when we want something more all-embracing and complete, perhaps even going so far (though this is by no means certain) as to make certain fundamental insights of biology fit in with the insights of mathematical physics. There is yet another problem, too, and this is a problem of words. We have learned to talk about probabilities, using the systems we manipulate at present. Perhaps we ought to examine our consciences on this subject. What is really meant by this word probability? It is a marvellously convenient word which puts us, so to speak, in the traditional swim, but which contains a fair number of traps for the unwary when we come to interpret our "know-how". This is the simple question I wish to leave with you.

Jean-Louis Destouches I think the debate on the fundamental unified field is more important than the one on determinism and indeterminism, and, like Mr. Vigier, I also think that if there is a fundamental group, it cannot be broken down, that is to say, we cannot consider it to be the direct product of two groups.

As to the question of determinism and indeterminism, this is a problem of language, of philosophy basically, but also a problem of technique. One can formulate theories classical in structure and having a hidden determinism, similar to the quantum theory. This is perfectly possible, it is just a matter of applying the technique of mathematical physics. But afterwards, it is up to everyone to make his own evaluation of the merit of such a scheme.

Von Neumann's theorem is obviously useful, though it has on occasion been misinterpreted. But let us suppose we have a new theory, this time reconciling the ideas of Vigier and Matveyev. The position of this theory will not be the same as that of the usual quantum theory. An analysis of its structure might perhaps show it to be intermediate between normal quantum theory and determinist theory — although a shade artificial, built up from hidden parameters — and then the opposition between the ideas of Vigier and Matveyev will not be as strong as it appeared to be *a priori*.

The Organization of Scientific Research

SYNTHESIS THROUGH ORGANIZATION OF SCIENTIFIC RESEARCH

Pierre Auger The mere association of the two expressions, scientific research and organization, raises grave problems, both cultural and social, spiritual and material. Many thinkers consider such an association positively dangerous: the spirit bloweth where it listeth and its visitations, insistences or renunciations are not to be organized; organize research, take away its complete liberty and it may wither and die. Yet contrary — and highly respectable — opinions are not wanting. La Rochefoucauld, for example, pointed out that it is not enough to possess great qualities, one must know how to order them. And order implies organization. Some will argue that such organization is an individual matter, a question of self-discipline, and that organization imposed from without must be harmful. This argument I consider fallacious. A man can organize his own gifts only within some broader social organization, a scientific and cultural tradition into which he fits — and here we notice at once that there are two aspects to this interaction between two forms of organization: one spiritual and the other material. The creative power brought into play in research is necessarily based on something acquired through education and communication, and it operates within an organized whole to extend, modify or even completely transform it. But even a thorough-going transformation has value only in relation to tradition; nothing can be reformed unless it already exists and is familiar. Many ideas, whose only fault was that they were too foreign to the pre-existing organization, have been utterly lost, or rediscovered long after they were first put forward — and then only if they had been recorded and preserved. Here, then, we already have two elements of great importance in good organization, whether of mind or of matter: first, a readiness to accept out-of-the-ordinary ideas and, if they cannot be integrated forthwith into the system, to preserve them. The problem is to discard the many unproductive trains of thought without risking the loss of something rare and precious.

> Patience, patience, patience dans l'azur
> Chaque atome de silence est la chance d'un fruit mùr.
> PAUL VALÉRY

"Patience, patience, patience in the empyrean Each atom of silence gives promise of ripe fruit", says Valéry. But one must be on the watch to receive the fruit when it falls and, if it is not yet ripe, be ready to put it in store until human thought and knowledge are ready for it. All this needs organization — not of the empyrean, but of those who gather the fruit.

The organization and ordering of man's higher gifts — those which confer upon him his supremacy among living creatures, his creativity, that is — thus lead inevitably to the dilemma: freedom or choice, for the individual as for society, and on the spiritual as well as the material level. The dilemma is whether to remain free or to make a choice which restricts freedom. The choice is based upon some existing organization which freedom can and does transcend. Man has to make this decision and, meditating upon himself, his past and his aspirations, to select the best course or, to use a jargon word, to "optimize" his decisions. But let us come back to the question of the practical organization of scientific research and see whether a synthesis is possible. It can be considered at three levels: the individual research workers; the institutional framework within which the work is carried out; and the material and financial considerations. Each of the levels at which organization is to be introduced again represents a single type of economy; thus it must be applied in a manner appropriate to the sort of problems handled during the research in question. In fundamental research, for example, the empyrean element, if I may use the expression, must claim the greater part, while in applied research, up to and including the development of final production processes for industry, the organization must be more and more detailed, the uses made of the fruits of human thought ever more disciplined. Thus, all I say about the men, the institutions and the material means must be adapted to the type of research proposed, whether fundamental, orientated, applied or operational.

As regards the men, we shall naturally be concerned with their quality and their numbers; the quality will be determined by their education and training — including character training — and by their personality and sense of vocation. Their numbers will depend on the stiffness of the selection procedures and the size of the pool from which they are drawn. It has often been said — and repeated — that science would advance more rapidly because the number of research workers was increasing exponentially; such an increase cannot continue indefinitely, it is bound to reach saturation point. This will be the point where the ratio of research scientists to ordinary citizens is optimum and where the numbers of both must either grow together or become stabilized. But we are still a long way from this optimum in the world as a whole and concerted efforts are still needed to train enough scientists, engineers and technicians. This is why it is so important to organize education, to maintain standards among teachers to ensure an excellent quality of recruits and to offer worthwhile careers to both teaching and research staff.

The institutional problem is too complex to be dealt with here in a few minutes. A full botanical treatise might be devoted to the various types of research establishment, ranging from the modest wild flower to the sap-filled forest giant and if, sometimes, that essential beauty which is scientific discovery emerges from the little flower, there are also sites where only great trees can flourish. Much has been said about the efficiency threshold of the laboratory or research institute. A lot depends upon the subject of the research, of course, but we must not forget that once a laboratory has crossed this threshold it has to be organized, even though the organization may involve a breaking down into smaller units, each enjoying considerable independence. But organization there must be, to nourish and protect those smaller units and keep them in touch with each other. If the subject of the research requires larger units, then it is at that level that synthesis will occur.

Finally — and it is here, perhaps, that the dilemma is most acute as between organization and freedom — there is the material and financial aspect. It is impossible to provide all research workers with everything they ask for; in any institute, whether private or public, national or international, there must be a choice made, priorities established, one request refused to allow another to be granted — all this, of course, done with immense tact and discretion, leaving great tracts free for the scientist, even at the risk of finding when the results are announced that mountains of funds and resources have brought forth a mouse. But optimization sometimes follows strange paths when the aim is, by its very nature, obscured by unknown factors; it may be close at hand or far away, there may be a broad avenue or a mere peephole opening on to the universe. And yet, scientific research remains one of the greatest of man's undertakings.

Can we, after this rapid analysis of the main aspects of organizational problems, come to any synthesizing conclusions? I think there are only three: one concerning the various scientific disciplines, the next the research workers themselves and their social and national origins, and the last what is now called science policy, operating on a national and international scale. The first call for synthesis arises from the growing importance of team work, work which cannot proceed without the formation of trained groups of research scientists from a variety of disciplines, both theoretical and experimental, because only groups of this kind can expect to succeed in some fields of modern science. We should not play down the organizational difficulties involved in setting up, directing and administering such groups; this is one of the scientific administrator's biggest problems. I shall quote a single example of complete success — the team at the Cavendish Laboratory in the days of J. J. Thompson and Lord Rutherford. A member of a team in a laboratory of this calibre, cracking difficult problems by the brilliant use of a range of assorted theoretical and experimental techniques, may indeed feel that he is a citizen of the Republic of Science. Meetings, discussions, seminars, visits and visitors seeking advice and encouragement, all these make life in such establishments very intensive, the chief risk run by the research scientist intellectually courted from all sides being that of having no time to get away by himself, and this is still the main ingredient of effective thought.

Of course, these teams whose importance I have been extolling transcend national barriers as well as disciplinary ones. It is often impossible to set up a group equipped to grapple with today's great problems without recruiting scientists of assorted nationalities. Such groups may work within national bodies, and delicate administrative questions may arise in consequence. But when the institutional framework itself extends beyond the frontiers of a single nation — for instance, we already have CERN and ESRO and we hope soon to have EMBO — the administrative difficulties become formidable; it is inadvisable to attempt anything like this except in certain well-defined areas of research, at least, with the world in its present state. These Round Table discussions are pointing up the exact reasons for the choices to be made. Science involves all mankind; equally it involves all man's faculties, the abstract and symbolic ones as much as manual dexterity and moral courage. It thus creates fine opportunities for men to feel deeply that they belong both to their own species and to the universe which gave birth to it.

Finally, and this is my third point, the new features of twentieth century science have required an increasingly definite intervention on the part of governments. This created the need to organize and administer government intervention which, in turn, led to the setting up of numerous advisory and policy-making bodies, instruments for consulting the scientists themselves, for consulting those who carry the political and financial responsibility, for making decisions about priorities, the selection of men, the establishment or reform of institutions, and the resources to be allocated to equipping and running them. This procedure is being adopted, established and implemented by all nations desirous of playing their part in the advancement of knowledge and of applying the results for the economic and social benefit of their people; thus science policy has stimulated international discussion, and some possibilities of synthesis are already discernible. Such discussions, leading to coordination, have long been a commonplace among scientists within their own "non-governmental" organizations, foremost among them the one which has brought us together here. Perhaps it will soon be time to attempt to make a synthesis of the syntheses; so far, the attempt has not seemed justified, moreover, the size of the task is rather daunting. But now that partial syntheses — of subjects, of nations, of methods — have been successfully achieved, a total synthesis begins to look less unrealistic. At a time when new links are being forged between the disciplines of science and these relationships are falling into a pattern which may one day constitute a single network uniting all the forms of scientific knowledge in the world — universe, life, thought itself — it would be a great pity if no parallel action were taken on an equally universal scale to make this human science belong to all men.

LESSONS THE HISTORY OF SCIENCE TEACHES

B. Kedrov Professor Auger who is a great specialist in the organization of science, has raised a number of very interesting points. The first might be stated as follows. The organization of science is itself a scientific task, it is even a new science that has developed in our time; it is concerned with science and its laws, its development, its particularities, the general and specific aspects of its various disciplines, and its dependence on historical evolution and economic factors. In a word, science is a subject for scientific research. I think Professor Auger is perfectly right in distinguishing between the spiritual and material aspects of this science of science: on the spiritual side, the theory of science, the logic, methodology and psychology of scientific research; on the material, the sociology of science, planning, organization and orientation.

My own feeling is, however, that the organization of science is not really a matter of construction but rather a question of avoiding the disruptions which, as a result of historical circumstances, hinder the progress of scientific development. Enlarging on Professor Auger's argument, I would distinguish four particular instances of this loss of continuity. First, the separation between science on one hand, and production, technology and practice, on the other. We all know the difficulties involved in the transition from scientific concept, even if worked out under laboratory conditions, to production process. Second, the break within science itself between the experimental and the theoretical. We often find either an accumulation of experimental data lying useless because of delays in publication or, conversely, conclusions reached by deduction awaiting verification by experiment. As far as the organization of science is concerned, it is true that contacts between both institutions and individuals do develop, between representatives of the natural sciences and representatives of technology, for example, or between the empirical sector of the natural sciences and the theoretical sector and mathematics. Then, I must mention the interdependence between those who primarily represent philosophy — integral knowledge and a broad understanding of the purposes of scientific research — and those who represent the individual disciplines (physics, chemistry, astronomy, mathematics, and so on). The break here is notorious. Historically, this is readily understandable, and it has in fact already been mentioned during the earlier discussion on cosmology. Cosmology was once the concern of philosophy, and its progress was almost non-existent. Now it has become a science in its own right and is beginning to move forward. But no problem of cosmology can be solved solely by the physicists, astrophysicists and mathematicians without the help of philosophy, without a comprehensive view of what, in the event, is the universe itself. The organization of science implies systematic working relations between philosophers and specialists in the natural sciences, so that general problems will not be divorced from those that belong to the specialized sectors. Fourthly, there is a particularly marked break between historians and the representatives of the modern natural sciences. The latter now all too often formulate their theories without referring back to the history of science, while the historians study science in antiquity, the Middle Ages or the seventeenth century and are scarcely concerned to bring their studies up to present times. Incidentally, may I say here that Professor Auger's introduction to the four-volume General History of Science, in preparation under the direction of Professor R. Taton, is a model of the way in which the historical study of science and the analysis of its present situation should be linked. If the two are linked, each reveals itself to the modern research worker in a completely new light. All modern problems are then seen in historical perspective; their origins provide a clue to future prospects, since the line of development is continuous, and past, present and future are logically interlinked in a single chain of events. The historians, on their side, will discover in the vast body of historical material, the germs of future concepts which, in two or three hundred years, or more, will become the theories of modern science. In other words, they will find those fruits Professor Auger referred to, which are not only capable of being transformed into new theories but may actually give birth to them, just as from human anatomy we can understand the anatomy of apes and prehistoric man, ancestors of the higher being that is man. I think it is precisely here that the organization of science may prove particularly fertile if it succeeds in organizing contacts, and possibly even setting up joint institutes (now only beginning to appear) where historical research in science would be associated with the study of the great fundamental problems of modern science. It seems to me extremely important that we are now realizing that theoretical research, and scientific research in general, are much more than a mere search for the best way of achieving a practical result; science must tackle its objective from every angle so that out of the thousand-and-one possible solutions will come one that is suitable for practical application. This is why the narrow utilitarianism which demands that all, or practically all, scientific research should have a practical outcome is so open to criticism. The range of science is infinitely greater than that of its feasible applications.

Team work, co-operation between specialists in different disciplines, interdisciplinary research — all these are commonly heard expressions nowadays. Taking modern research in the life sciences, for example, we find that, through the analysis of the living organism, classical biology and biochemistry provide certain chemical data but, alone, can furnish no reply to the problem of the essence of life; for that, co-operation is necessary between experts who, so far, have had very few contacts — biologists, chemists, physicists, biophysicists, cyberneticians, biocyberneticians and mathematicians. A synthesis must be made that takes in the different aspects of life, so far studied separately. And it is here that Einstein's example of scientific synthesis is particularly important, in regard not only to ideas but also to practical organization: that is, bringing together men who often find each other's language practically unintelligible (the biologist, for example, and the mathematician or cybernetician), and seeing to it that these specialists in various disciplines not only understand one another but actually join forces to tackle the common problem. Just as Einstein solved the problem of the relations between matter, movement, space, time, mass and energy by a single theory that covered them all, we in turn must attempt synthesis; and here the problem is still more complex.

Robinson Crusoe would not get very far in present-day science; the results of individualism, as opposed to team research, will be scrappy and incomplete. On the other hand the team must not destroy the scientist's individuality or shackle his freedom; it must foster his creative spirit, and this again is a question of organization, of enabling the scientist working as a member of an interdisciplinary research team to give all he is potentially capable of giving. Here even the most minute details may be of great importance.

I observe in conclusion that we have completely ignored research in the field of psychology, yet there are a whole series of problems awaiting investigation which might enable us to understand the creative process in the scientist's mind. Just one example: Mendeleev. The famous inventor of the periodic table devised it in a highly original way. One of his favourite pastimes was playing solitaire, and as a result he developed an extraordinary faculty of patience. When he came to arrange his sixty-three elements in a single system, he tackled the problem like a game of solitaire with cards on which he had noted the known elements and set about combining them; and this is how he solved his problem. The study of the mysterious, unique course of thought that eventually leads to truth is an exceptionally important scientific task. It is one that already involves research in the field of psychology and it merits all our attention.

P. Auger Mr. Kedrov's final comments on the way in which scientific creation occurs seem to me to be particularly stimulating. He mentioned the example of Mendeleev; there are also other cases in which the writings and statements of research scientists show us how they made their discoveries. Maxwell, for instance, published a number of articles in which one can see his theory of the electromagnetic field slowly taking shape. He subsequently removed all the scaffolding which he had initially used to reach his famous equations. For this, he relied on a series of fairly rudimentary and, in the early stages, almost clumsy models which were gradually perfected and which led to the concept of interconnected electric and magnetic fields and their variations as represented by his equations. The mathematical model for Maxwell's equations is thus the outcome of a series of more or less material models which recall Mendeleev's game of solitaire to which Mr. Kedrov alluded. A thorough study of scientists and discoveries about which accurate historical information is available would certainly be of great value in elucidating somewhat more clearly the psychology of the creative worker or, at any rate, the scientific creative worker. This is a problem of organization, but of internal organization, the form of organization which I mentioned at the beginning of my remarks: organization by the scientist of his own thoughts and his own method so as to advance towards scientific discovery. A well-organized mind is of course more likely to succeed than a vague one, speculating at random. Even if such minds occasionally capture an idea in passing — as a butterfly may be caught in a net — they will certainly not attain the efficiency of those minds which systematically follow clearly delineated paths and use very extensive nets in order to capture such new ideas as may emerge from the Brownian movement of thought that goes on in the brains of all of us.

BOTTLE-NECKS AND INTERACTIONS BETWEEN DISCIPLINES

P. Piganiol It is not easy to speak after what we have just heard. I personally was particularly impressed by the references to the internal structure of science, and the role of scientific history in the development of mind and thought. However, I think that there are a few rather down-to-earth observations which might be added at this point. To begin with, a certain typology of thought models is being gradually worked out, and I think it can be said that much of the progress made in chemistry, both pure and applied, rests on the fact that chemists have established stocks of models on which they can draw when doing research. Chemistry is probably ideal for this, since it is so easy in chemistry to discover simple models to show how our thinking evolves. But what I want to demonstrate is that the problem very soon arises at two separate levels once one leaves the individual discipline and seeks to assist individual persons. They can be helped in two ways: by being shown the structure of the links between disciplines, or by being provided with results. I shall explain by two examples. The first, the question of links between the disciplines, is a delicate matter. Within a given discipline, thought models may have developed (significant progress, for example, in chemistry, chemical engineering and kinetics) but we are far from clear regarding the links between disciplines and the advance necessary in one to enable another to progress. When it comes to organizing research — which too often, unfortunately, boils down to distributing the funds — we should know where the bottle-necks are (for example, why extend a highly specialized biological chemistry if there is no protein chemistry at all?). Here, very little has been done, but things are beginning to happen and, like Professor Kedrov, I very much hope that the matter will be thought about and studied. There is also another way of facilitating interaction between disciplines, and that is not to limit the exploring of a given concept to a single discipline only, and to explore, systematically, series of things which at first glance might not seem to justify doing so. I will take a Soviet example that has always struck me because it is of the kind that often intrigues the West. In chemistry, our Soviet colleagues seem to make a pastime of establishing every conceivable diagram of the very numerous mineral salt solutions. Every month they publish vast quantities of data that might at first sight seem to have no interest at all; but this is to ignore the underlying scientific strategy which is unquestionably valid. The purpose, obviously, is not an advance in chemistry itself. The arsenal of data will no doubt serve for some applied research or the exploitation of a salt deposit in a dried lake in southern Russia; as far as application is concerned, these systematic results are certainly of value. But this scarcely suffices to explain why our Soviet colleagues take them so seriously when a few experiments would clear up the variation or variations on any particular practical problem and its solution. In fact, however, they go into everything very thoroughly. A Soviet colleague replied one day to my question that, by definition, there are many mixtures of salts in sediments in nature; it is therefore worth providing future geologists with an extremely detailed picture, sound and irrefutable, of the way in which all saline deposits successively form. For the moment, the little experience we have is quite insufficient as a basis for geologists to make any important progress in their own discipline; and naturally enough they have no desire to undertake a task they regard as being outside their field. What then is the answer? The job is routine but delicate, and it is educational. Why not have it done, therefore, by young people, who will thus be trained as they work? This alone will make it worth while, since training, in itself, is beyond price; and, at the same time, data will be accumulated which will very likely be of service to geologists.

Here a very clear distinction is visible between two levels at which two disciplines interact. At the first, concepts interact, so that the progress of a concept in one may transform another, for example, infra-red, ultra-violet, harmonic resonance and their associated mathematics enabled organic chemistry to make extraordinary progress. In addition to this interaction of concepts, there is a more commonplace one: the interaction of results, an effect produced by masses of data — which, sometimes, one must have the courage to accumulate. Here, the vital decision is administrative, and decision it is: the scientist who makes it may sometimes experience a feeling of regret, because it takes him away from immediate concerns which, in regard to his own particular branch, may be richer and more promising. At this level, science involves solidarity within the discipline, solidarity of effort, and a certain humility.

In these problems of research, I feel we must always remember that although science can sometimes be extremely exciting and profoundly satisfactory in a way that arouses our eagerness to continue, there are also enormous tracts that are dull but essential. To accept this philosophically, we need solidarity, the full human solidarity that the pursuit of knowledge implies. That is all I wanted to add on this point, for I was much impressed by Mr. Kedrov's thought-provoking remarks.

P. Auger When a research scientist in a laboratory is asked to spend many years on a tough assignment that offers no major results but is none the less vital, then I feel that it is simply good organization to allow him at the same time at least a taste of the scientist's paradise. During part of his time he should be free to do as he pleases, anything he rightly or wrongly considers more creative, more directly constructive and spiritually satisfactory. In a large laboratory in which various scientists have tasks that are frequently tedious or difficult, and being precise demand the utmost attention if they are to be of value (because the work has to be done, a matter of human solidarity, as Piganiol has just said), then I believe that spiritually and psychologically sound administration requires that they also be allowed to do something else as well — they should have not exactly a hobby, but some opening to let them feel that, indispensable as their everyday task is, they can get away from it ever so often. Frequently they may fail, but they will have the satisfaction of being able to try, of having at least exercised the creative faculties and capacities which all of us possess, even if it be only a spark.

THE ORGANIZATION OF SCIENTIFIC WORK AMONG THE SCIENCES AND IN RELATION TO TECHNOLOGY AND CULTURE

G. Holton It seems to me that a consensus is emerging among the members of this panel. With respect to the organization of scientific work, I see three general topics: organization of scientific work itself; science and technology; and science and culture.

The first is the easiest to discuss. The problems of scientific organization are much in the news and much with us — partly because they are complex, but partly because they can be solved. One always likes to deal first with the solvable problems. Among them are these: How shall we apportion the few dollars we have among the ever-growing numbers of scientists, some of whom in the rapidly growing communities of scientists want all of it? In my field, physics, there were about fifty physicists in the U.S.A. in 1900, and last year there were 27,000 members of the various branches of the American Institute of Physics! In other scientific fields, the rate of increase is not very different.

But these numbers are not quite so frightening if one looks beyond them. Thus, the relative number of professionals outside science has also grown in the same general way. The fraction of scientists among trained professionals has been almost constant over the past fifty years. Secondly, the problems of communication among the very best, at the top of the pyramid, are not at all as severe as those for the people near the base of the pyramid. The people at the top of each pyramid need to communicate with only a few other people standing near them or at tops of other pyramids. And this has always been so. A major contributor at the time of Galileo knew and kept up with the work of as many really creative scientists as does a correspondingly good man in physics now. Information-retrieval is not a very serious problem at or near the pinnacle of creative work. The problem lies below that point; it is, however, one price we must pay for encouraging so many people who cannot do first-rate work. Of course, they are needed! They support the work of the first-raters at the present level of large-scale activity. This is a social decision we have made, and I see no reason to regret it. But though we should try to solve the information-retrieval problem, we should realize that it will not materially affect the creative lives of the very best people.

Nor do they, many of these, in most Western countries, now lack the essential means for their important work. Pierre Curie — in 1902, I think — was told by Paul Appell that he would be a candidate for the Légion d'honneur. To this, Curie replied: "I do not feel the slightest need of being decorated, but I am in the greatest need of a laboratory". At that point he had only the wooden shed at Rue Lhomond and two small rooms at Rue Cuvier. This is not happening any more. The needs will always be larger than what is available, but a basic, decent support of the best people is in sight — and where it is not, it is perhaps merely the fault of the scientists themselves for not taking the strongest and most courageous action against bureaucrats who stand in the way of reasonable support.

A much more difficult problem than whether or how to support science is the question of which science to support. What share should be given to, say, physics, and how shall we apportion resources among the very expensive branches and the inexpensive ones within a science? In the United States just now we are going through a crisis concerning this very problem: for example, many outstanding chemists believe they are getting far too little of the total science research budget. It is quite possible that within basic research the decision for support in the future will be guided more by the claim and promise for social usefulness and need of the fruits of research in the long run. It is obvious that this component in a total policy will be difficult and dangerous to implement. But some additional criteria will be needed soon, if only for this reason. As everyone who has done original research knows, all the sciences are, as it were, interacting in a symbiotic metabolic process (particularly so at their mutual borders) so that the advancement in any one science really makes sense from a long range point of view only if this advance is not at the expense of other sciences. There are many examples to show that the solutions of important scientific and technical problems have kept back the progress of physics research because we did not know enough about metallurgy, or about chemistry, or about mathematics. I might say parenthetically that what is true for the different parts of science is true for all parts of the world of scholarship: for any to function healthily, they all must flourish.

The next problem that I address myself to very briefly is science and technology. Here, we are in a very much worse plight. Technology today — at least in certain Western countries, and to some degree everywhere else also — is a blind giant being propelled by three great forces. One is the industrial private profit motive, the second is the ambition of the military, and the third is State planning on the basis of narrow, nationalistic politics. I am afraid that these three do not suffice in the long run for a sound direction for technology or for the uses made of science in technology. These three forces are also tugging to a certain on the center of gravity of basic scientific research itself, but in technological fields they are the major influence, and as a result one can see that many important and acknowledged needs are not being adequately met. (The need to share technical resources and expertize with the less developed countries is only one of these.) To become masters of this giant, we must program into the progress of technology both a statesmanlike and a philosophical component, analogous to those which we can now find in the case of science itself.

Thirdly — in even worse condition — is the relationship between science and culture. For example, the rapid advance of scientific ideas is leaving many people behind who feel that their psychological balance itself is being threatened by their ignorance about the essential truths. In 1908 my teacher, P. W. Bridgman, who later received the Nobel Prize for his work on the physics of high pressures, entitled his thesis: "Mercury resistance as a pressure gauge". It was a very classical, beautiful work — and he could have told undergraduate students in their first year at university all about it, whether they were going to be scientists or not. This is of course no longer possible. No longer can one tell all one's students or even one's colleagues what is being done at frontiers, and the lay persons, particularly the most intelligent ones, feel this gap very strongly. A greater and greater gap between the citizenry and the scientific elite is worrisome not only because the scientists must not lose contact with their citizen-supporters, those who must pay for the work, must furnish the future scientists, and must intelligently decide about policy involving science. I am worried also because I believe that unless the intellectuals know or could know, at least in general terms, what is going on at the frontiers of knowledge in all fields, including science, they are caught in an impossible and paralysing ignorance which has no precedent in the past.

Lastly, a particular point involving Unesco. In this beautiful city, in the middle of the 1920's, there was formed by the League of Nations an Institute of Intellectual Co-operation. Curie, Bergson, Einstein and many others were members. Something like this again is needed today. The questions I speak of are not going to be solved by those who engage themselves primarily or exclusively in purely scientific tasks. These inquiries must be carried out among scientists and scholars who feel they must also give time and efforts to discussions and planning in which each can, for the moment, shed his national identity and his loyalty to a small area of expertize. I hope that Unesco will make forums of this sort in the future.

The time is both urgent and appropriate. I see a new answer emerging to the question we must always be ready to ask again — the question why advanced societies should support science and its organization at all. In the seventeenth and eigteenth centuries the answer was that each genial man and woman ought to do that which his beneficent daemon tells him to do — at least as long as he does not get in conflict with authority. In the nineteenth century, the reason for organizing and supporting science was primarily that our industrial society needed the by-products of science. For the twentieth and twenty-first centuries I suggest the answer is different: the answer will be that now society in the developed areas of the world has the obligation to organize toward the satisfaction of a much neglected human right, *the pursuit of a satisfactory career for every one of its citizens in any field that the citizen regards as being meaningful to him and which is not socially harmful.* Applied to science, the organizing principle is to make it possible for everyone who wishes to study and contribute to science to do so.

Our societies should no longer be preoccupied with an organization that is primarily a device for battling over ideological differences; it should no longer be preoccupied with the provision of elementary wants because technology can take care of this. The real task, I would suggest, for the next fifty years or more, in the organization of science as in all other pursuits, is to assure more truly a human use of human being.

P. Auger The question of physicists and biologists or chemists has been raised and, as has just been pointed out, it has caused plenty of ink to flow, not only on the other side of the Atlantic but also here in Europe. Physicists, particularly if concerned with high-energy particles — the nuclear physicists — need instruments that become more and more costly, and more and more men to handle them. For some years now we have also had the space physicists, perhaps still more demanding in terms of finance, resources and manpower, since a vast number of people are needed to prepare the launching of instruments, animals and, above all, men into space. The question is likely to become more and more acute in the years ahead. It is very well stated in a book by Fred Hoyle, just published, containing the texts of three lectures, which contrasts the dinosaurs of the physicists — the enormous instruments they claim to be indispensable — and the small laboratories of the biologists, biochemists and their many neighbouring specialists. Personally, I find the point a little curious, coming from an astronomer, since a hundred years ago it was the astronomers who had the dinosaurs: they alone had huge, extremely expensive instruments, each almost unique of its kind. But it is true that this situation raises serious problems, especially if, after space science, another — one that wanted to explore the depths of the earth, for example — should in turn demand large sums and manpower resources; or other sciences or branches of science should come into existence and become similarly demanding. So far, I do not think that the economy of States is being affected — they can bear the burden without being seriously embarrassed. I do not wish to raise the more political question of whether national defence can derive enough from such activities to justify devoting a proportion of its own resources to them. There is, for many States, a national problem of organizing research, and also an international problem. Bodies such as CERN (European Organization for Nuclear Research), which was called into being here, in Unesco, are asking a lot from Europe, especially if the proposed new 300 giga electron volt accelerator is built, involving expenditure that is beginning to worry scientists working in other branches of research. We must not blink the facts, but look things frankly in the face. Round-table meetings — or triangular, if you prefer, with space specialists, nuclear specialists and biologists — might get somewhere as far as the psychology of the scientists themselves is concerned. As was pointed out earlier, advances in one direction in a discipline link up with advances in others, and there is no doubt that many of the results physicists have obtained, with their great machines, concerning isotopes, radiation and so on, are useful and have greatly helped biologists. We are thus advancing, not in extended order, but jointly. Some branches, obviously, demand more and advance more rapidly, but we must look at the whole, and I feel a frank and honest discussion would be most helpful. Unesco might well be the place for it and I may perhaps recall what was said a moment ago: Unesco provides a major intellectual forum within the United Nations, and many operations like those we have been discussing might well be included in its program. Mention might also be made, incidentally, of the alleged opposition between science and culture which has occasioned so much discussion since Lord Snow's famous study of "The Two Cultures". The debate already existed, but now has its own specific characteristics. "The Two Cultures" has become an accepted phrase, but we are very far from having solved the problems it involves. I feel that one solution might be to ensure that an educated man — the *honnête homme* of the eighteenth century — will have learned enough about science to be able, without strain, to read reviews and keep up with modern science instead of simply exclaiming "How amazing!" as his jaw drops.

RESEARCH IN THE HUMAN SCIENCES

M. Debeauvais The organization of research also involves problems in the social sciences which may be considered at two different levels. First, what are the criteria for the most effective organization? From this point of view, the various arguments already put forward can be applied also to the social and human sciences; here too, it is becoming increasingly possible to supervise the development of knowledge through the systematic organization of research. This has special implications for the social sciences, since organization problems are coming more and more within social science terms of reference: administrative science, sociology of organizations, and so on. *Inter alia,* there is a permanent conflict, only now beginning to be analysed, between a trend in all systems and societies towards more and more complex and oppressive organizational forms, and the necessary innovation which is all the time in danger of being crushed out of existence by organization. Fostering, or protecting, innovation represents a new and difficult problem on which the social sciences can certainly cast some light.

The effectiveness of research may be increased and the progress of knowledge speeded up by combining various disciplines in interdisciplinary research — not only the natural sciences but also the social and human. This raises problems of communication — between individuals, and between disciplines that are at very different stages of development. No one any longer thinks a hierarchy of sciences — as in Auguste Comte's famous pyramid — possible. It is, however, a fact that the scientific disciplines are at very different stages of development, and this raises extremely serious problems of communication, not only between one discipline and another but even between different social sciences, for example, the difficulty economists have in communicating with sociologists. These problems of communication are also within the terms of reference of the social sciences and branches of social and of individual psychology are at present studying them (theory of information, etc.). Here again, the social sciences, as has been pointed out, are needed to ensure the general progress of human knowledge.

It is being increasingly recognized, in practice as in theory, that the scientific approach, hitherto primarily used in the natural sciences and biology, may also be applied to man and human societies. This implies an extension of man's power over himself and over human societies. With this in mind, one may well deplore (as specialists in the social sciences frequently do) the overwhelming disproportion in terms of material resources and research workers available to the natural and the social sciences respectively; or even, following Comte, erect a pyramid of the credits allotted, by disciplines, for research in all countries, or of research workers, classified by discipline — either is a valid measure of the range of research. It will be found that, the more a subject is remote from and outside man, the greater the range of the research and the resources devoted to it.

Considering the matter from the point of view of equality between disciplines, this is not particularly serious, since certain sciences will inevitably be more developed than others and have more extensive resources. It may be more disturbing if we go somewhat deeper, since aims are not irrelevant: the sciences concerned with the organization of human societies or with knowledge of man are doubtless more important from an anthropocentric viewpoint than knowledge of the structure of matter. In any case, the consequences of decisions regarding the organization of human societies are more important than those that follow decisions on other subjects and, when one knows that the progress of knowledge now largely depends on the human and material means devoted to a branch or discipline, the problem of the organization of research can be placed on an altogether different level: how in fact are priorities established when organizing research — because the priorities are implicit, even if science is accepted as one and indivisible? Theoretically, all disciplines are equal, all scientists are entitled to the same status and even the same honors. In fact, decisions are reached in quite a different way, and not by scientists but by governments and, in my view, power is the main objective that may be attributed to the decisions of governments trying to orient the progress of knowledge. I am not endeavoring to offer a value judgement as to the nature of these aims; I merely observe that scientific progress as such does not at present decide how research as a whole shall be organized. It may obviously be asked whether this is inevitable, desirable, or outside the scope of our discussions here. My feeling is that it is absolutely central to them, and that there is no point in talking about a synthesis of the sciences if we refuse to consider it.

It is far from easy to try to justify priorities. Criteria independent of ideologies might perhaps be sought, for example, priority for problems that affect the greatest numbers — one not easy to apply, but an attempt might be made. Decisions affecting the increasing gulf now recognized to exist between the underdeveloped and the advanced countries will vitally influence the future of human society, but this problem receives no priority treatment in deciding the scientific policy of governments, and research on it is very far from being proportionate to its importance.

Another possible criterion is the instinct of self-preservation, the survival of the human race. Yet if we consider the gigantic efforts at present being devoted to the natural sciences (which are vital and fascinating) and thereby to the structure of matter, it is apparent that what primarily interests governments and what underlies the priority given to these sciences is the destructive power they confer. Societies (and the human sciences) are no doubt trying to make behavior a little more rational, but only within the very limited frontiers of individual nations, whereas national frontiers and a rationality limited to the nation are insignificant beside investigations of the structure of the atom and the possible consequences: perhaps the annihilation of mankind.

If we now try to take a census of those research workers studying relations between national societies and the vital problem of progress and innovation in international relations which could save us from nuclear disaster, we find that they come at the very bottom of the scale of priorities in the organization of research.

Similarly, our glimpses of knowledge about man himself or his organization in restricted groups (sociology, psychology, and so on) are likewise power oriented, and more used to manipulate men and societies than for their emancipation. The resources allotted to research are again involved here, at any rate in part. We need only compare the number of psychologists and sociologists paid by enterprises which are certainly not aimed at the destruction of mankind but do aim at conditioning individuals (market surveys, advertising, and so on) with the number of research workers specializing in disciplines which have no utilitarian function. Here again, the disproportion is overwhelming, even within the human sciences. I regard these fundamental problems — which Gerald Holton raised and discussed better than I could do — as being of capital importance. It is not enough just to be optimistic, and to note that there is an increasing interest in science and that science is developing more and more rapidly.

P. Auger The applause which has greeted your remarks, Mr. Debeauvais, indicates that there are many here who are directly intersted in the human sciences. Even if they are in fact specialists in the natural sciences, they are very clearly aware of the great importance of these other sciences, and realize that the inadequacy of their development at present — an inadequacy which is perhaps not entirely due to lack of resources — has created a disequilibrium in the management of human affairs. The possession of substantial resources — means of destruction but also means of production — along with very little understanding of how to use them and of the results they may have from the human point of view, obviously leads to disequilibrium, and there I feel you are entirely right. What I do not see quite so clearly is how the human sciences can develop better and much more quickly solely by obtaining greater resources. It is possible, but I am not sure that it is only a question of resources; I think there is a certain growing process in each science and that a period occurs, that a point is reached, when these sciences take wing, if I may so express myself, or take off, as they say in industry. It is possible that a number of the human sciences have not yet done so. It is only when they do that resources should be provided in very considerable quantity to enable them to keep flying and not, as unfortunately happens sometimes to aircraft that take off too soon, crash to the ground.

P. Piganiol The question which has just been raised is a serious one, regarded from the angle of human societies. Within these societies there are men who decide, who lead, who act, and there is the society itself, which may well wish to have some say in its own destiny but does not always succeed in dowing so. As regards scientific policy, something quite remarkable has happened in the United States: gradually, as the White House put together the building bricks of a scientific policy, Congress set up its own very extensive parallel study and information machinery and services. As far as I know, this is the only case outside Eastern Europe (where the purpose is the same) of society producing its machinery for discussion and selection rather than scientific or governmental circles, where the power factor was perhaps too frequently the dominant factor in decision-making. And this is a problem which we are going to meet, one in any case that we cannot avoid: as scientific society develops and our research strategy and organization grow more confident, society as a whole must have its say. As things stand, it need hardly be pointed out, our traditional parliamentary structures no longer have the slightest power to question or analyse. This is highly regrettable, and one of the very serious problems of the modern world.

P. Auger This seems to me to be very directly connected with science and culture, the problem referred to by Gerald Holton. By and large, our parliamentarians are educated, but not in science. This may largely explain why scientific committees set up in parliaments to examine the effects of the government's scientific policy on society are not always too successful; the interest is kept up for a few months, and then it dies a natural death mostly, I think, because our parliamentarians do not know enough about science to have any passionate interest or see where it is all leading. This is unfortunate because, as Mr. Piganiol very rightly suggested, the executive has no real intermediary between it and the people: on one side, the scientists who belong to the people and study social problems as members of various groups and, on the other, the executive. The legislature is supposed to provide the link between the two, but in fact does not, in France at any rate, and in many other countries. Yet in England, the House of Commons has a very live scientific committee. I have personally attended its meetings on several occasions and it seems to me to play an important role. In the United States, the organization is still better and, there again, I have been personally able to attend the discussions of certain Congressional bodies. These constructively criticize the actions of the Government, on the basis of a sound knowledge of the facts. The result is either support or, conversely, proposals for reform or change. Every government should keep in touch through the elected representatives with the nation's vital forces — this indeed should be the normal practice in any democracy.

Vladimir Kourganoff I do not think the problem of the organization of research can be considered in isolation from that of its impact on the organization and training of people to do research. This problem is always being swept under the carpet. There is increasing rivalry, amounting almost to antagonism, between the claims of higher education and those of research. Having been indiscreet enough to write a little book entitled *La Recherche Scientifique* *. I can hardly be accused of underrating the importance of research. But in the third and last edition of my book, motivated by what I consider to be a continuing scandal, I have had to change several paragraphs to make it quite clear that research will die a natural death if it is to remain a mere psychological priority, a prestige matter, a sort of frill — which is what it has come to be for many government agencies and, sad to say, for many university people, too. They must be made to see that this is a dilemma of the same kind as that we have already been told about between individual research and organization, only here the dilemma is between the amount of effort to be put into research or into teaching in the widest sense, ranging from the training of research scientists through the dissemination of scientific knowledge to the consolidation of the known facts. It is essential to realize that the ratio of teaching to research is not simply an individual drama acted out by each university professor on his own account, but an urgent problem which must be faced at institutional level and hence a vital part of the problem we are discussing here.

Once it is accepted that university preferment depends exclusively, or almost so, on research successes, any devotion to teaching chores will be doomed in the long run. No professor with young assistants working under him can fail to see this and be disturbed by it. It is a harrowing drama. For a long time there was no problem here, because the assistants did not belong to a research team and where able to combine their teaching activities with research. But once they become part of a research team, in physics, for example, or astronomy, teamwork becomes paramount; they cannot drop the assembly of a piece of equipment being built by their team to set a problem for their advanced students, to discuss their work with them, to organize a seminar, or even to write a popular science article, much less attempt a synthesis of newly acquired knowledge. Conferences like the present one, and Unesco itself, must acknowledge this problem and face up to it. Let me put it this way: the organization of research employs research scientists and they have to be trained by somebody. Therefore people must be trained to train them. Above all, there is no question of a rough and ready training such as formerly sufficed, when a man went, for example, to the Cavendish Laboratory, worked as

* Collection "Que sais-je?" Presses Universitaires de France, Paris.

apprentice under a master like Rutherford and in due course became a master himself. This artisan-type or self-help training is no longer adequate. We must organize training and consequently we must devise some sort of division of duties. Those who have a bent for teaching should concentrate on teaching and not be thought any the less of if, at a certain point in their carrers, they ease up on research and devote themselves to teaching.

P. Auger This point Professor Kourganoff has just raised concerning research and teaching is absolutely fundamental in any country seeking the harmonious development of science. There is more to this than seeing that men who concentrate on teaching will have good enough career prospects without being tempted to neglect teaching in favor of research; research may offer more rapid advancement, perhaps a seat in the Academy and various other distinctions. There is also a type of research which does not pay off in great discoveries. The scientist who works in obscurity, who does a useful job conscientiously, unrewarded by brilliant discoveries leading to the social advantages I have mentioned, ought also to have a worthwhile career with some opportunity to exercise his imagination. Here again, I think there ought to be some sharing, but there will not be unless we set out to see fair play. It is not just a question of half-measures and palliatives to meet the immediate difficulties which will allow us to have teachers and train students in the same old way, even though this may not be the most efficient one. There is another problem, too, mentioned recently by Nobel prizewinner, Professor Jacques Monod. He said that in his own discipline, molecular biology, higher education had been very slow in getting off the ground, and was still not properly organized, whereas botany and zoology were still being taught over a number of years. This situation is gradually being put right but there are still many universities where things are even more backward; we cannot be complacent about such a situation. It is particularly bad in France, but it may be less so in other countries where the universities are not so much under government control and many have been able to establish chairs and set up courses to match the development of new branches of science.

Reverend P. Leroy The Chairman referred earlier to the research worker's choice of discipline and research. I should like to mention another problem, namely, the selection of research workers. Some laboratories give the impression that the young men and women working in them are more or less amateur, and this is because they do not have deep down the research worker's basic sense of mission, which is a very personal thing. It demands great intellectual honesty, a lot of imagination, an inventive spirit, but also the habit of observation and reflection. It demands an analytic mind as well as a feeling for synthesis. These are qualities that cannot be simulated; hence the responsibility of educators and those who select the research workers. Judging by what happens in our countries at any rate, candidates are proposed by research leaders or directors on the basis of work done. This is a very inadequate criterion. Many young men and women prepare theses with an enthusiasm they very quickly lose, and end up by occupying the post but not playing the part.

A second, far-reaching problem that may be outside our terms of reference here is information. Excellent abstracts are available — I have in mind especially the CNRS *Bulletin signalétique* which we use daily and which is well produced — but, overwhelmed as we are by the number of publications devoted to the minutest speciality, perhaps something new in the line of general organization might be attempted; this would affect not only research workers but also governments, if anything worthwhile is to be achieved.

P. Auger You have certainly brought up two problems of great importance that I personally had scarcely alluded to. The choice of men, of research workers, has so far followed traditional methods: competitions, examinations, personal recommendations by those who know the candidates. These methods work more or less well, but could certainly be greatly improved. One major improvement would be to expand the reserve to which I referred earlier. It is not enough to have gifted young people; they must be sought out, because they are very often hidden. I mean that there are young men and women who are very gifted but neither they nor their professors realize it. There should be ways of enabling them to give what they are capable of giving. We should not only select from among those who apply but, by suitable publicity, stimulate vocations and offer opportunities to those who do not yet realize they have the necessary qualities. Television and radio have interesting possibilities here. Books like the "Que Sais-je?" series are also very effective, I feel, in arousing the interest and attracting people capable of rendering great services who would not normally be led, in the social patterns in which they and their families live, to embark on this type of career or take up this kind of work. Once again, organization is needed.

The second problem, information, is immense. We all know the avalanche, the tidal wave of science publications — greatly varying in merit, incidentally. According to recent statistics it seems that many of them are read by only one person in the ten years following publication, while others are read by hundreds of thousands. This variability in usefulness is really excessive. However, they must all be cataloged and stored. As I said at the outset, it happens that brilliant ideas are not immediately understood or their potentialities appreciated; they must be preserved, perhaps to be taken up again later. Such was the case of Gregor Mendel, who was understood, not in his own lifetime but forty years later. But his work had been published and so survived. Hence we should publish practically everything, but not allow ourselves to be swamped. The average research scientist, who cannot spend all his time reading, must be able to find what he needs with the help perhaps of automatic techniques. This is almost as serious in sciences as the problem of population growth for the world, with its attendant food and other major difficulties. Publications may become so numerous that they no longer serve a purpose, are no longer read, can no longer be utilized; duplication or triplication could become common in research, the same task being undertaken in many places simply because no one knows it is being done elsewhere. This is undoubtedly a major problem, and one that should receive attention.

G. Holton I shall add a brief word. In the United States, education is the biggest "industry" in the country. Forty-two thousand million dollars last year were spent in one way or another on education. Over 25 per cent of the population is engaged in education as teachers, or as full- or part-time students. And yet this whole enterprise is largely being conducted with the intellectual and technical tools of the nineteenth century, and is heavily overladen with ritual that is more ancient still. The main reason why so many people teach so badly and others learn so little, or why teaching does not have the *éclat* that many say it should, is not difficult to find: teaching is difficult, and little is known on a fundamental level about learning. Taking refuge from these problems, many of our teachers, even in the sciences, rely on folklore, for example, that all information must come out of a book. Instead of making the learning experience an immensely exciting affair, with all the aids and drama of films, of real encounters in the laboratories and the like, they essentially read a dull book to the student and except the book to be reconstructed during the examination.

Worse still, many teachers have not thought through the essential conceptual structure and philosophical basis of the subject. Consequently, very often — in the sciences particularly — they teach a mere catalog of information.

To produce remedies, we need, in every region, a lively research center on teaching and learning, a Saclay of learning acids. If science itself were being pursued the way teaching is today, we would not have discovered very much! We must decide to make learning and teaching much more exciting and effective, and we must be ready to invest money and manpower for this cause.

B. Kedrov The problem is extremely important, and seems to me to be linked, not to any particular national conditions, but to the whole historical background of our present age. The real issue is the maintenance and organization of scientific schools.

If teaching and research are separate, the student entering a scientific institute will already have a trained mind; whereas it is in the interests of specialized schools to develop vocations for specialized studies young, for example, at the university or even at secondary school.

Let us take an example in another domain. We have spoken here of science and culture; let us consider how a painter or musician is trained. The career of a student who has come young under the influence of a great painter or composer will be quite different from one whose youth has been spent under more humdrum conditions, with no opportuniy to develop his talents.

We, too, have this problem in the Soviet Union, and I would like to tell you about an experiment we are making in the training of mathematicians where, as you know, intellectual training must begin very young — if a mathematician is not trained by the age of about twenty, he will never be. Hence our leading mathematicians have organized, under the direction of Academician Kolmogorov, a special boarding school for children selected for their mathematical ability from all over the country, even the remotest parts. Kolmogorov and other leading specialists supervise the studies of these children, who are trained to take up higher mathematics, becoming specialists in turn and eventually the colleagues of their former teachers. At the moment, this is only an experiment. I have visited the school myself. The children are extremely interesting and yet in a sense they are no longer children. Three years before they finish the secondary course they are already studying things like the theory of relativity which — I have checked this — belong to the third year of the university syllabus. In other respects, they are mere children, their minds not yet trained; but in mathematics, their training begins earlier. This obviously creates its own problems, including the danger of overspecialization.

In short, the questions which have been so rationally considered here call for the closest attention. The results of experiments everywhere should be assembled for examination and analysis, and here Unesco could play a very important part. In any case it is, I think, quite clear that only under the working conditions of research laboratories, and in contact with outstanding scientists who do both teaching and research, will the great scientists of tomorrow be trained.

Reverend F. Russo Professor Auger has spoken of the dilemma as between freedom and organization. This obviously lies at the heart of the problem of developing research, but I wonder whether we are not gradually overcoming it, succeeding in promoting mutual aid between organization and freedom and bringing to an end the mistrust which prevails between the creative individual and the organizer and which, in my view, is one of the sources of difficulty hampering the progress and development of research. I think we must understand that research today can no longer be left, as it was in the past, to the play of obscure, spontaneous forces; we have embarked on a conscious and deliberate process of research development. This may be rather a broad and philosophical view of the matter but I think it should be kept more closely in mind, constituting the basic drive animating the determined organization of research. Those currently entrusted with this task should overcome their fear of being accused of being technocrats; they must adjust their sights and appreciate that a lively and fruitful association between freedom and organization is perfectly feasible.

This brings me to a question raised by Professor Kedrov, namely the role of history in the organization of research. I am inclined to share his views, though with some reservations. I believe that only modern science, the science which immediately preceded contemporary science, is useful for the purpose of research organization. For instance, what has been said here about Einstein shows that it would certainly be profitable to think over his work again and that this might help us to handle both cosmological problems and the tricky subject of indeterminism; but I do not think it would help us much to go back any further. We should not learn much about organizing research today from a study of Kepler, however interesting such a study may be in a different context. The main reason why we should gain little profit from a study of the past is that until recently science was spontaneous and unorganized, in contrast to what it is today.

I should also like to support in a very general way Professor Kedrov's hope that we shall see a science of science develop. I hope his appeal will be widely noted, for we have no such science of science at present. Perhaps I could suggest two lines of research: first a pure science of science and later an applied form which would guide us in the conduct of research. These two fields, as I see it, are not quite the same thing. In the strict sense of the term, I should define the science of science as the effort we must make to understand the broad movement of science and the links which exist between the different sciences. I have been surprised to notice that, even in the present forum, these matters have been referred to within a frame of thought which is still very empirical; it has been said that there is a connection between this and that piece of research, for instance, that the isotopes which have enabled such remarkable progress to be made in biology are a pleasing by-product of nuclear research. This is quite simply accepted as a fact, but we now have to go beyond this and concern ourselves with the structure of science, both its present structure and its dynamics. We already have some of the elements needed for such an investigation; science as practised today is already structured, and far be it from me to suggest that scientists are not aware of the relationships that exist between the various disciplines. What I do suggest is that we could develop research much more surely and efficiently if we applied ourselves more deliberately, systematically and scientifically to studying the structure of science. This would involve, for example, doing away with that blinkered view which sees relationships only in the form of the classic family tree — those traditional subdivisions which as structures are totally inadequate. I am astounded that, with such a range of mathematical structures available to us, we are content to retain these crude structures when dealing with the relations between the sciences. There is a great deal of work to be done here, for instance, to determine the local, regional and ultimately the overall structure of a given field of science. This would enable us to distinguish various levels within the structure of science, and work of this sort would be intrinsically rewarding. We have been talking of the structure of science in relation to the needs of research organization, which makes the whole thing seem rather utilitarian and prevents us from grasping the full significance of the structure of science and all that it could stand for.

We should then, I think, have to develop a science of the conduct of research which would undoubtedly enable us to make use of what we had learnt about the science of science, and we should have to try to translate the actual structure of science into the structures of research. This, however, is only a part of the answer, for the organization of research will never be determined solely by the structure of science. There are extra-scientific considerations: when we have to determine what proportion of a nation's total budget should be allocated to research, or when we have to decide between the claims of pure and applied research, we are going beyond the purview of science.

Here I venture to say a word on the comments made earlier concerning certain research objectives which lie outside the scope of science. Mention has been made of power. Power, it seems to me, is unquestionably one of the criteria governing the choice to be made in scientific research; but if we look closely at the present behavior of those responsible for the organization of research, I think we shall have to modify our views. The various criteria — power, affluence, comfort, prestige (and prestige is not the same thing as power), even the advancement of knowledge — are not, in fact, independent of one another, they are closely interrelated, and I do not think it is fair to say that today the nations are seeking power alone. Reference has been made to the very considerable sums allocated to education budgets, and these budgets are not designed to contribute solely and directly to the power of nations. Accordingly, as I see it, a more qualified view should be taken of this matter. The fact remains, however, that power objectives do still preponderate to an excessive degree; and, as has very rightly been said, it is regrettable, indeed, disgraceful that policies affecting scientific research should not pay more heed to the needs of the developing countries. At the same time, it should be mentioned that a United Nations conference held in Geneva in 1963, and in which Unesco participated directly, did raise this question and, in my opinion, helped to foster a greater awareness of these needs in scientific circles. In considering the scientific conduct of research, we should stress the extent to which research is conditioned by factors external to it and, in particular, the part played by information. It is distressing that scientific information should not have been tackled with the required energy at government level, except perhaps in the United States. In Europe, in any event, we have a situation which is all the more deplorable because there is no lack of either money or manpower; the missing factor is that some governmental or scientific body should have taken proper cognizance of these needs.

In conclusion I should like to make a few brief remarks of a more theoretical kind on the criteria which should govern choice in research. Of course, research is by definition an advance towards the unknown and we cannot determine beforehand what will be found. But I believe this to be rather an over-simplified view and, in line with the work in psychology which has been quoted here — not to mention studies in logic, we should chart the relationships between disciplines. Scientists might be able to see them more clearly if they were to temper their freedom with an effort at logical understanding of interdisciplinary relationships, as great research scientists have often done, in fact. Another factor which can help to make research less indeterminate is the fact that its spontaneous dynamism tends to be self-orientating. I should like to refer to work which was done here, under the direction of Professor Auger, with the object of determining trends in research and of defining a coherent research policy on the basis of a knowledge of those trends. Finally, I would add that the conduct of research tends increasingly in a number of sectors towards a convergence of effort which constitutes a powerful aggregate — take NASA, for example, or research coordinated at the international level, like that with which Unesco is concerned in the fields of hydrology and oceanography. These great aggregates should be considered much more systematically. Here we have an organization, a structuralization, which is already remarkable but which might be more consciously and purposefully shaped.

P. Auger I shall not take advantage of my position as chairman in order to comment, point by point, on all Father Russo's remarks; but I would like to mention a few. For a start, I think it is worth spending time on the historical study of outstanding men, like Galileo. It is fascinating to analyze the intellectual make-up of someone who could well serve as a model for a good few of our research scientists, and I think such a study would justify the time spent on it.

Then, as regards the science of science, are we not confusing this with what is properly scientific philosophy, a form of research one might suggest to those of a philosophical turn of mind who, having already or wishing to round off an adequate scientific education, would be willing to devote many hours, years even, to working out such a science of science from a purely intellectual point of view. It would be a sort of complete philosophy of science, to be distinguished very clearly, as Father Russo indeed did, from the art of the conduct of research. I do not call it the science of research, as this is rather different. Complementary to the art of the conduct of research would be the true philosophy of science, and this is what I consider the science of science, providing an understanding of the unity between the various parts of science which is now becoming manifest all around us.

Science, indeed, is presenting a united front. We are accused of knowing more and more about less and less, and in one sense this is true, but only as regards our individual work. When we look at the whole range of scientific interest open to any one of us, we see that on the contrary this is much wider. There is hardly a research scientist of my acquaintance, be he physicist or biologist, who does not maintain a broad interest in all the other sciences, even though he may be working on a very specialized problem. One has to specialize these days to do effective work and make fresh discoveries. These scientists already discern in the grand panorama of modern science the emergence of a single pattern embracing the whole range of human knowledge — the universe, life, even thought itself.

I agree with what Father Russo has to say about power; prestige, however, is a part of power, maybe one of its better parts and not too subject to considerations of military might or world economic influence. I think prestige is often no bad counsellor. This was certainly the motive which inspired certain governments of the past to encourage science at a time when it had no direct influence on national defence or economic development. For example, astronomy, which I was talking about in a previous session, was always the spoiled child of governments in the seventeenth and eighteenth centuries and even into the nineteenth. Nobody argued about astronomy and the astronomers were always granted the large sums they asked for. Why was this? Because, for prestige reasons, states were keen to have their own astronomers so they could claim their share of successes in this spectacular if not notably useful science.

F. le Lionnais A myth is beginning to spread among the general public that some discoveries are entirely collective and that individual research is of no account in them. All of us here know that this is not true, but I think the general public should also be brought to realize the fact. In many cases, it seems to me, collective research, carried out by a team, is aimed in the first place at creating a climate, at developing a mutual stimulation and an exchange of information, between members of the same research team. This is of course extremely valuable, and more and more necessary nowadays, but I feel that the really original element in research still remains individual. This applies, in my opinion, to all the sciences, and probably to none in so high a degree as to mathematics. Non-mathematicians belonging to an elite — a non-mathematical elite — often ask me what theorems were discovered by Nicolas Bourbaki. I have regretfully to answer that Nicolas Bourbaki discovered nothing in mathematics comparable to the theorem of Pythagoras, or Fuchsian function, or even Sturm's theorem to which, at the end of the nineteenth century, Professor Sturm referred in his lectures as "the theorem whose name I have honor to bear". No, Bourbaki discovered nothing comparable, but the did create a very important mathematical climate, without which it would not, I think, be possible to understand the time in which we live. Some recent mathematical discoveries could not be explained if their authors had not studied Bourbaki's *Eléments de Mathématique* and discussed it among themselves. I should not like to give the impression of criticizing this renowned group, which includes in its present membership some of the world's greatest mathematicians (three of them, for instance — Schwartz, Serre and Thom — have been awarded three of the eight field medals which, as you know, represent the Nobel Prize for mathematics). But their work, and the work of others whom I shall not list since there are so many outstanding mathematicians in this group, is strictly individual. I think this is a point worth making.

Still on the subject of mathematics, I should like to revert to Mr. Debeauvais's idea — which seems to me very interesting as a way of throwing light on the dynamism of the sciences — of classifying the sciences according to what they cost, their budget. The cheapest science of all is, as it happens, mathematics. It is a curious fact that mathematics costs less not only than physics or chemistry, but also than the human sciences, archaeology, history, and even comparative literature for which electronic computers are now being used, for as you know, although mathematicians worked out the principles of computing, they do not themselves use computers. This surely suggests that we should reflect once more on the fundamental nature of mathematics. I have just returned from a visit to the Soviet Union, where I was greatly impressed by what they call the Mathematical Olympics, which allow for an increase in the number of students from among whom future research workers may be drawn. At the beginning I thought that these Olympics were rather like our *Concours général* in France. That is so. But what has happened to those who won first prizes in the *Concours général?* How many of them have taken up scientific careers? Unlike the *Concours général,* the Mathematical Olympics as organized in the Soviet Union are followed by fellowships, by a great many fellowships, so that ultimately the country's scientific potential is increased.

Lastly, may I add that Father Russo has already expressed almost exactly what I wanted to say, and expressed it much better than I could have done. But I should just like to revert to Professor Auger's remark: I agree with his distinction between the science of science and the science of research; but I do not agree with the expression "the art of research", at any rate in as far as it has an exclusive connotation. Over and above the present art of research, I should like to see a science of research established, and I do not feel this to be purely Utopian. I am persuaded that in the structures and methods of research we can find structures more or less comparable to those of mathematics. Given two methods, we may ask whether they can be added together or merged; we can consider at what points they intersect or complement one another; or we may even ask whether we can multiply a method by something or whether they can be classified in a "group", in a certain order. Such work really goes beyond the scope of the present round-table discussion. I have referred to it because it seems to me to justify, at least as an objective, the term "science of research".

P. Auger Mr. Le Lionnais has pointed out that mathematics is the cheapest science. But I should like to ask him, to begin with, whether mathematics is in fact a science. This is not absolutely certain. I think it differs from all the other sciences in such a way that it cannot really be compared; in other words, it has to be treated separately. Comparison of costs would not, therefore, have a great deal of meaning; nor, indeed, would comparison of teams, since, although teams do exist in mathematics, they have there neither the same value nor the same importance as in the experimental sciences, at any rate in those possessing an important experimental basis. As regards information on mathematics likely to arouse an interest in mathematical careers, I should like to mention that Christophe's *L'Idée fixe du savant Cosinus* is again available in the "Livre de Poche" series, and I think that this publication could play an important part in increasing the number of young people wishing to take up mathematics.

Mr. Le Lionnais also mentioned that in the Soviet Union, from which he has just returned, there are Olympics, enabling mathematical gifts to be systematically discovered. I have seen the same thing in the United States. The same systematic search for talent exists there, and what is interesting is that the progress of young people "discovered" in this way is followed. Note has been taken of the number of those who in fact chose the careers for which they had been selected, as it were, by this sort of *Concours général* which is organized on a very broad basis each year. The results have proved to be not unsatisfactory, although they were, of course, far from reaching one hundred per cent. A good many of the young people selected in this way — that is to say, recognized as having a gift for research — have subsequently taken up interesting careers, though there may not be an Einstein among them. What is involved, therefore, is not just a game but a mechanism which has a real influence over the development, within society, of that intellectual elite of scientific research workers which is becoming more and more necessary, as each of today's speakers has acknowledged.

M. Debeauvais I should like to say just one word, Mr. Chairman, concerning the objection which you and several other speakers have raised concerning what I said, and which I consider to be entirely justified. Even if additional resources were awarded to the social sciences, this would not mean that they would yield immediate results as spectacular as those we are now witnessing in the physical or natural sciences. But this is not the angle from which I wished to view the question. It is probably inevitable that the social and human sciences should be much less advanced than the natural sciences, because their subject matter is of such complexity that we are only just beginning to glimpse the way in which the scientific concepts developed in the natural sciences might provide some keys that would let the social sciences into the pre-scientific age. Viewing the question from another angle, however — that of degrees of urgency (of survival or of a humanized society) — I think it is important to make a great drive for research in the social sciences, even though it may not be very effectual in the short term, so as to remedy a fundamental disparity which may have the most serious consequences for mankind.

P. Auger This is very true and although some, if not all, the social sciences are still in the proto-historic stage and have not yet achieved take-off, there is no doubt that a greater effort, more money and above all more people, could speed up their development considerably. If prehistoric man in the neolithic age had been given the means for artificial development, provided with new materials and the means of subsistence so that he did not have to spend all his time hunting food, in short, if he had been able to enjoy a little leisure, he might have advanced more quickly than he did. Similarly, the social sciences might have a better chance if they were given more means, if more people took up their study and also perhaps if they were to make more use of the various information media to let the general public, and young people in particular, know what is going on in the social sciences. Many students with cultural interests choose to study literature and never even consider the social sciences. They know what the exact sciences and the natural sciences involve, mathematics and technology, and they think, "That wont do for me, I'm no good at maths, so I shall take classics, French, literature or the arts", and they reject the social sciences, simply because they fail to appreciate that these, too, can offer a most interesting career. It is a scientific career with a great future, although it does not require so much in the way of mathematical ability. We have an interest in seeing that humanity develops in a more balanced fashion, rather than rushing precipitately and almost blindly down the primrose path of technology.

I think it is time to bring these discussions to a close. It may well be that we have not sufficiently emphasized those points on which we are all in complete agreement. That scientific research should be organized is not only a good thing but a necessary one at the present time, for governments, for the international groupings to which they belong, for the research laboratories and institutions, and for all of us who have to do with science as, when our work is organized, we can make the best use of such talents as we may possess.

PART THREE

Teilhard de Chardin

Knowledge of Nature and Man

TEILHARD DE CHARDIN AND EVOLUTIONISM

Jean Piveteau The central theme of Teilhard's scientific thinking was evolutionism — not just biological, but extended to take in the whole of reality. I shall try to explain to you in a rapid review of his work how he approached this problem, and the form in which he presented his coherent synthesis.

If we follow the course pursued by his thought, we shall see how such a synthesis was gradually built up.

The first palaeontological work Teilhard did was on the mammals of the early Tertiary. From this contact with primitive forms of life which, despite their lack of morphological definition, somehow foreshadow their descendents, he draws the fundamental conclusion that life is *flexible* and *progressive*. Being flexible, it is liable to change; being progressive, its evolution does not proceed in a random manner but is orientated. As to the direction of orientation, Teilhard concluded at the end of a paper devoted to primitive forms of primates that the line favoring the largest brain was the most successful and thus represented a privileged axis of evolution.

In a number of essays Teilhard explains the meaning and scope of the evolutionist conception. It is a special way of studying organized beings: the historical method. The historical perspective is not merely something applied to the study of life, it is a form proper to the human mind. Increasingly our experimentally based science is tending, in its investigations as in its constructions, to adopt the historical method, that is, the evolutionary point of view. "One of the strangest phenomena to occur in the field of thought for half a century," wrote Teilhard, "is unquestionably the gradual, irresistible invasion of physical chemistry by history." The life sciences had already been mobilized in this movement.

The mysterious necessity compelling this invasion was the discovery of time or, more precisely, of the true meaning of time. "Time", Bergson was fond of saying, "is what stops everything from happening all at once." Reality did not spring up all at once in its present form but was built up little by little; we do not live in a finished Cosmos — we are involved in cosmogenesis. In the moving words of the Apostle Paul, "The things of this world pass away."

All the time Teilhard was investigating fossil mammals, another theme of meditation was maturing within him and was soon to preoccupy him almost continually. This was the problem of man. Although he neither discovered nor studied fossil man, Teilhard's ideas on the subject were to give a fresh orientation to our thoughts on human palaeontology. In a paper, probably written in 1922, and unknown except through lengthy extracts published by Edouard Le Roy in his fine book, *Les origines humaines et l'évolution de l'intelligence,* Teilhard set out the exact nature of the problem of the relationship between man and nature.

For naturalists and philosophers alike, man had always been an equivocal creature — *Homo duplex,* Buffon called him. Some would relegate him to the animal kingdom, ignoring or misrepresenting his unique features, while others made of him a Universe, set in opposition to all but his own kind, tossing in splendid isolation upon the vast waters of the world. Teilhard's great contribution was to preserve the uniqueness of the human phenomenon without plucking it out of the frame of familiar experience. The only way to do this, he declared, was to create a super-category to express the fact that man, however strong his links with the onward march of life, marks the end of one phase of development and the beginning of an entirely new one. At the origin of this new entity Teilhard conceived of a special sort of transformation acting upon pre-existent life; he called this humanization and later defined it as the verifiable fact of the entry into this world of the ability to reflect and to think. Since then, the significance of the human phenomenon and of the factors effecting humanization have become two of the main avenues of research in human palaeontology.

At the same time we see the emergence of that essential characteristic of Teilhard's method, restoring life and man to the cosmic perspective. Man's relationship with life is equivalent to that seen to exist between life and matter. Vitalization of matter, humanization of life, these are two events which dominate the history of the Cosmos.

Teilhard went off to China in 1923 on a trip intended to be relatively short; in fact, he was hardly to leave Asia for over 20 years. While keeping an eye on the discoveries concerning Sinanthropus, that early hominid contemporary with the deposition of the red soils, he was also working as both geologist and palaeontologist. But, most important of all, his ideas about man were gathering strength: man, who cannot be set apart from life; life, intimately bound to the Earth.

Thus, when he explained the great phenomenon of the expansion of the continents by granitization, his mind was constantly haunted by the problem of man. Without the Earth, would there be man, and without man, what would Earth be? The final purpose of a science still to be created and which he proposed calling geobiology would be to establish the relationship between the two evolutions which took place in geological times: the evolution of the continents and the evolution of life, or as he called them, continentalization and cerebralization.

In the years preceding World War II, and particularly during the time he could not leave Peking because of the hostilities, Teilhard could not carry on his field work and this slowed down his laboratory work. But his meditations continued unceasingly and he was then able to work out the synthesis sketched in his earlier essays. What kind of evolution does he now propose to us?

His starting point is man. Teilhard picked up Pascal's theme of man's anguish as being most appropriate to show man's frailty, straying between the two abysses of immensity and nothingness. "Let man consider what he is as against what is; let him see himself as lost in this small, out-of-the-way province of nature; from this mean dungeon wherein he dwells — I mean the Universe — let him learn to appreciate at their true worth the earth, its kingdoms, its cities and himself." Man, life, seems such a very little thing in the infinity of worlds. We should also remember the words of the astronomer, Sir James Jeans, quoted by Teilhard. "To what are life and man reduced? Fallen, as it were by mistake, into a universe obviously not made for life, we remain clinging to a fragment of a grain of sand until cold and death return us to crude matter; we strut around for a little while upon a tiny stage, knowing full well that all our aspirations are doomed to failure in the end and that all we have achieved will perish with our breed, leaving the universe as though we had never been. The universe is indifferent, even hostile, to all forms of life."

Can such an outlook be scientifically acceptable? Can we really hope to understand the universe, life and man if we start out from the geometrical perspective of immensity? Let us, as Teilhard invites us to do, look at things from the biochemical aspect of their complexity — a third infinite — and there occurs within our field of vision a complete reversal of values, as though we were looking through the wrong end of the telescope.

This parameter of complexity enables us to establish a genetic classification of the material realities which have appeared in the course of time. In this way, one can draw up a table comprising simple bodies at the bottom, above these molecules, then viruses, next living cells and so on. The order of complexity corresponds to their chronological order of emergence. Now let us look at the universe from this perspective. Take the biggest objects for a start — the *nebulae*. Their substance is extremely tenuous and is probably just hydrogen, that is, the simplest element of matter: one nucleus plus one electron.

Going up one step in immensity, we look at the *stars*. They are much more complex compared with the nebulae. But stars do not go beyond a certain stage; they are the laboratories where nature manufactures *atoms* from primeval hydrogen — nothing more.

It is on the *dark planets* alone — and nowhere else — that we have the opportunity to pursue this mysterious ascent of the world towards higher complexities and where the evolutionary effort devoted principally to the manufacture of large molecules will henceforth be concentrated.

Among these planets, at any rate in our solar system, Earth is probably the only one which is life-bearing. It may be the sole center where the synthesis of large molecules goes on, and the living creatures which inhabit it probably represent "the most complex compositions to emerge from planetary geochemistry".

Thus, life no longer looks like a dirty stain upon some tiny point in space but appears as "the specific effect of the complexification of matter".

Although the parameter of increasing complexity enables us to find a meaning in the evolution of the universe before life appeared, it will not serve, Teilhard tells us, to determine the direction of evolution after the appearance of life. How can we allocate comparative complexities to a plant or an animal, an insect or a vertebrate, a reptile or a mammal? We have to look for another parameter.

The more complex a living organism, the more self-centered it becomes and at the same time its self-awareness increases. Such a form of complexity is proper to the organ directly controlling perception, which is the brain. While it cannot be assessed by its content of atoms or molecules, it can be evaluated by its level of organization.

Palaeontology shows us a progression in the encephalon right through geological times, from the Primary to the present. "Among the infinite modalities throughout which vital complexity is dispersed", writes Teilhard, "the differentiation of the nerve substance stand out as a significant transformation. It lends a meaning to evolution and subsequently proves that it is directional".

And man? He is the ultimate expression of a line which was individualized by the primates in the jungle; at a geologically recent date, as estimated by Teilhard, but in very far-off times, as it seems to us, an entirely new phase in history of life was initiated. Man is capable of meditative thought; alone among living creatures, he is endowed with the power to put his mind into a higher gear and to perform analysis. Every human consciousness has the faculty of auto-consultation and hence is able to discern the conditions and consequences of its actions and then, to some extent at least, to guide them accordingly.

Thus the processes of life are always directed towards the realization of the richest and most differentiated nervous system, the genesis of a better brain which will be the instrument of a further developed, freer and more integrated mentality. Of course, there are failures and regressions along the way, but the general direction is clearly defined. "Everywhere else except in man perception has reached a dead end; in man alone has it continued to advance."

So, today, man stands at the summit of psychological complexity. True, he can no longer look upon himself as the center of the world, this being an ancient idea which has no meaning except in a static universe. But better than this, he is the arrowhead of evolution, and he it is who will change its course and shape. In its infra-human phase, evolution has been characterized by ramification and divergence; no doubt, this was even more the case in the earliest eras of prehistory. With modern man, it turns back on itself as though it has reached its final flowering. Here we have convergence, not divergence.

While still in this infra-human phase, living things worked unconsciously, as though by inertia, towards the general advance of life. So far, life had nothing to fear from its creatures.

But along with man was born the faculty of judging and criticizing reality. What does he think about this impulse which produced him, from which he is issued? Is it worth going on with? He cannot dodge these questions, and they signal the appearance of a moral sense in the world.

Thus, in Teilhard's brand of evolutionism, the universe and life represent the story of progressive complexification, crowned by the flowering of conscience.

To conclude this rapid survey of Teilhard's ideas about evolution, I should like to make two comments of my own.

Can this upward thrust from the simplest forms of matter to the human spirit be identified with a monistic outlook?

In reality, the evolutionary series as reconstructed by Teilhard represents *thresholds.* Although the complexity of matter is defined by the level of organization of atoms or molecules, the complexity of life is expressed by the development of a moral sense through which man becomes capable of self-criticism. There is thus a threshold between matter and life and another between life and man. This does not necessarily imply discontinuity: new qualities can quite well emerge through a generative continuity of evolution.

Teilhard's achievement, the peak of his scientific thought in our eyes, was to integrate man into evolution and to make clear to us the true significance of the phenomenon of man.

Man is a key to the past, but a key to the future, too, for in him we see revealed the flow of the primeval genetic processes, and he represents their final impulse; he raises to a new power where all is made explicit the energies of the universe.

And so, with Teilhard, and thanks to him, we rediscover man's unique greatness, as proclaimed by Antigone:

> The wonders of the world are many,
> But the greatest wonder of them all is man.

Reverend P. Leroy As I have been asked to describe Father Teilhard's scientific career in a few words, I shall limit myself to a quick account of the disciplines to which he devoted more than 50 years of his life: geology, mammalian palaeontology and anthropology. In these very different fields his intuitive yet at the same time critical mind moved along lines both simultaneous and converging. The previous speaker has told us that Teilhard's ruling idea was to discover from a study of the past the guiding thread to man's future. In his research, carried on with consistency and method, he was able thanks to a program which was at once flexible and rigid to enrich his current studies by the knowledge which others had acquired. His strict scientific training, whatever we may think about it, underpinned the new and constructive character of his synthesis, exhibiting the multiple planes of his expertise and placing for us man and his destiny.

Within the frame of pure science, the breadth of his vision enabled him to introduce original insights which are now well on the way to becoming classical. In his capacity of geologist, he — better than any of his predecessors — was able to decipher the daunting complexities of the stratigraphic levels of Northern China. With his stubbornness and perseverance, he made comprehensible the "Chinese Shield", that vast deposit of terrigeneous and lacustrine formations covering the Great Central Plain which before then had been poorly understood. From this time on, the geological strata of the early Tertiary and the Quaternary form a harmonious and coherent whole. Moreover, the "Old Chinese Plinth", which had mystified the geologists of the early twentieth century, became less mysterious. It was in the course of his research on ancient mineral masses that Teilhard conceived and developed a new theory of granitization. What did this mean? The classical theory of the transformation of rocks and sediments into granite by metamorphism, that is, by the combined action of heat and pressure, did not seem to him to account for the presence of granite in the middle of recent strata. Thus, the idea of the intrusion of granitic masses into younger rock formations began to take shape in his mind. "Sial" would no longer be, as formerly proposed, a virtually invariable mass, but would increase in a constant manner. The continental masses would slowly grow by the addition of new granitic material from below. Teilhard considered that this phenomenon of granitization was what had built China. He found the proof of this in the granitic intrusions of the Upper Permian, the Jurassic and the Upper Wealdian on the borders of Mongolia, in Santung, Tsinling and so on. A similar observation had been made in the Miocene of the Japanese archipelago, that is to say, in the middle of the Tertiary. Teilhard was not content merely to formulate a new theory, he moved on to generalization. Continental drift was a myth; Wegener's theory did not take into account the findings of geologists; Asia was not moving away from America; Asia was "neither drifting nor contracting, but expanding". The universe was no longer to be thought of as an inert mass which had been at rest for numberless aeons but was modifed by accidental changes. On the contrary, the Earth moves, it is alive and subject to important factors which govern the expansion of the mineral masses of which its crust consists.

Teilhard devoted a large part of his life to palaeontology. Specializing as he did in fossil mammals, he was particularly interested in the primitive fauna of Europe and especially of France; in China he was concerned with fossils of more recent times. Whatever category of forms he studied, all his reports carry the feeling of the need to "come up for air" so as to be able to understand the whole story. Pure anatomical descriptions, research on affiliations or analogies between the various groups, these things were indispensable but they never seemed enough to him. His was not a mind to be satisfied with fragmentary data — he always wanted to go to the heart of the question; for an intelligence of his type, this seemed the most reliable way. Thanks to his powerful intellectual bent for grand synthesis, he understood evolution with unusual force. For him it was no interplay of genetic combinations, forming in a random manner from the potentialities which every living creature carried within itself, a condition to which it is often reduced, but a vast movement embracing the whole of existence and obeying a law of recurrence which is manifested throughout the processes which link "material things" to "reflective processes". Looked at in this way, evolution overflows the area in which many scholars wish to contain it. Doubtless, everyone agrees that the evolution of life should be seen as a linking of successive forward steps and an uninterrupted series of new forms. But there is a danger of remaining too submissive to the systematic views of transformism as a theoretical vision, or of a genetics unable to account for all that we know about the unfurling of life. Teilhard takes hold of the question at a higher level; without being able to give a rigorous explanation, he sees evolution as "something cosmic", over the whole universe the evolutionary laws, which have been experimentally proved, must be repeated at every level.

Father Teilhard's orientation towards palaeontology was not a matter of chance. From the time he was twenty, the problem of man's origin seemed to him to be fundamental. He could not accept that man was an intruder in the world of living things or that he could have found his place there without everything being prepared for it from a remote past. Anthropogenesis cannot separated from cosmogenesis. Circumstance allowed Teilhard to be in the right place at the right time, to be at the very core of the most important anthropological discoveries of the last fifty years. First there was the find in 1923, in the Ordos desert south of the great southern bend of the Yellow River, of prehistoric settlements, reminiscent of the Mousterian or Aurignacian industries of Europe. This discovery, made in the company and under the direction of his colleague, Father Licent, threw fresh light on our knowledge of prehistoric man in Southern Asia. Shortly after this, Teilhard was present in Peking when the remains of an undeniably human fossil were brought out of the collapsed cave ol Chou-k'ou-tien and were named Sinanthropus or Peking Man. The geological study of fossiliferous strata later enabled Teilhard to take part in anthropological research in South Africa where Australopithecus was being excavated. In the meantime in Southern China he had established the synchronism of the layers containing Sinanthropus and those containing orang-outangs. In Burma he made an inventory of an ancient Palaeolithic site; in Java, which he visited twice, he saw strata containing Pithecanthropus. There can be very few scholars who have the good fortune to visit so many sites which nurtured the ancestors of the human race.

The results of all this travelling about the world were recorded in prolific writings; some were purely scientific reports but others were philosophical essays stimulated by his findings, and among these there eventually appeared *The Human Phenomenon*, which he had long conceived and meditated upon. He was still working at the Wenner Gren Foundation for Anthropological Studies in New York when death overtook him on April 10th, 1955.

This, very briefly, was the distinguished scientific career of this great thinker. One can, it is true, criticize his extrapolations as too bold, or his ideals as too pure. But one cannot criticize his contribution to science; this is a definite fact which no one can deny. Father Teilhard de Chardin, Member of the Académie des Sciences de Paris, was a true and outstanding scholar.

O. Costa de Beauregard I had not been thinking of speaking in this debate. As you are aware, physicists are usually very shy of going outside their own discipline. I am thus a little scared of the idea leaving familiar ground to speak in a debate on Teilhard de Chardin.

I do so because I am directly interested in problems connected with the irreversibility of time, problems I have discussed in my book, *Le second principe de la science du temps.* These are cosmological problems of a kind a physicist may set himself, and indeed, many physicists throughout the world are doing just this at the present time. Unquestionably such problems are closely interrelated with those raised, first by Bergson, and later by Father Teilhard.

You are, of course, aware, that physics adopts a fairly definite position on the irreversibility of time. The first manifestation of this way of looking at things is enshrined in the second law of thermodynamics, according to which physical systems evolve towards a state of uniformity from which no usable energy can be derived. Very briefly, as I do not want to go into the details, this is a theory which says that the universe is moving towards a state of cosmic death.

These problems were subsequently revived by statistical mechanics when it was realized what was concealed behind the notion of entropy — entropy being a physical value which always evolves in the same direction, growing larger by definition. Behind these problems of thermodynamic irreversibility there was also a problem of statistical irreversibility, this being the tendency of inert matter to revert to its most probable states. Boltzmann, in particular, established an equivalence between entropy and the logarithm of probability.

In terms of probability, the problem arises in the following form: what is the underlying reason which makes the universe tend towards the most probable states in the future, and why is it that, on the contrary, it does not seek more probable states when they are situated in the past? This is a question which appears paradoxical but to which there is no simple answer, because it is not comprehended in the actual formalism of the calculation of probabilities. When we examine this question, I think we fall back on the conclusion formerly reached by Van der Waals in an article on irreversibility, the same conclusion I have had to fall back on myself, as have many other researchers all over the world in one way or another: there is no intrinsic irreversibility in statistics, but there is an extrinsic irreversibility which is introduced by an *ad hoc* principle, Bayes' theorem, which says that, in calculating inverse probability, extrinsic coefficients selected *ad hoc* must be introduced into the problem.

When we try to understand the expectations which motivate this judgement, this "No Entry" sign in front of blind inversion, we see that we are led by physical considerations to relate the system under study to ever larger systems and finally to consider the whole universe. When we reach this stage, we are no longer quite sure what we are talking about, as constructing a statistical system to fit the entire universe is a somewhat rash undertaking! But one thing is certain, when one looks for the real root of physical irreversibility, one finishes up at a level which is undeniably cosmic.

This problem relates to other physical problems, particularly the problem of the irreversibility of radiation. We all know how waves are divergent or retarded: when a stone is thrown into a pond, we observe a series of waves diverging, going out from the stone and dissipating its energy. But no one has ever seen a series of waves converging upon a stone to eject it into the hand of a passing stroller.

Now a certain number of workers have recently had the idea that there is a relationship between these two principles, the growth of entropy and the retardation of waves. The philosophy which emerges from this, as I see it, is that behind the various principles of physical irreversibility lies one ultimate principle. I have spoken of wave theory, but in reality all waves are quantized, so that we can reduce the problem of the retardation of waves to a statistical one — I have proved this, and so have other people — situated within the framework of quantized waves. So that, finally, all these principles of physical irreversibility come down — in my opinion as in that of other people — to a single ultimate principle of irreversibility which could perhaps be, in a certain formulation, a temporal application of Bayes' theorem. This is one aspect of the matter.

But there is another. I told you how thinking about irreversibility in physics leads one to a cosmic perspective; one is also led, just as inevitably, I think, to a psychological perspective — and this is where cybernetics comes into the picture. Cybernetics introduces from the word go the play of the idea of information, which can also be defined as the logarithm of probability. And now that the theory of cybernetics has been fairly well established, we can see that this idea of information was implicit in all the classical problems of the calculation of probability, and particularly in the problems which Pascal and Fermat set themselves in the seventeenth century, and on the basis of which they founded probability calculation. Cybernetics has shown up something we had no idea of before, namely that when information is acquired by means of observation, physical experiments or, more generally, by any other common act of everyday life, such information is necessarily acquired at the expense of a fall in the negentropy of the environment. As Gabor wittily put it, you do not get anything for nothing, not even information.

This quite new and revolutionary insight introduces a previously unsuspected link between the content of perception and the whole of the so-called material universe. Cybernetics has thus introduced a concept which had not been noticed before between that passive physical process of evolution described by Carnot, evolution in the direction of increasing entropy, and that psychological process I must also classify as relatively passive, the acquisition of information or knowledge by observation.

But at the same time cybernetics raised another question — I am speaking about general, abstract cybernetics, not the technical sort used by designers of machinery — and this is the reciprocal question of the conversion of information into negentropy. This has been much debated in books on the theory of cybernetics, in particular by Brillouin, and the entire analysis of Maxwell's daemon is made on this basis. When a certain piece of information is acquired about a physical system, we at once have the power to put it macroscopically in order. There is the possibility of converting the information into negentropy. This has resulted rather oddly in the unpremeditated rediscovery of the old Aristotelian ideas. Aristotle thought information had a dual aspect: it could be on the one side the acquisition of knowledge and on the other the power of action, the power to put in order, or organize.

In this connection, too, we must ask ourselves about what Mehlberg in the U.S.A., calls "law-like irreversibility" and "fact-like irreversibility" of time. This is what I was talking about a little while ago, on the subject of Bayes' theorem, but the problem crops up in cybernetics, too, the question being why it is that the process of conversion of negentropy into information, in other words, the process of observation, and the acquisition of information, is so much more straightforward and simple than the inverse process of converting information back into negentropy, that is, the process of action.

I do not think one can give a categorical answer to this: it is the state of being in the world; for us, observation is more restful than action.

Entropy is converted into thermodynamic terms by means of a coefficient of conversion, the constant k. The acquisition of knowledge is very cheap in terms of negentropy, but to produce negentropy is very expensive in terms of information/action; this is the situation in which we find ourselves and I think it is directly described by the smallness of the constant k.

Just as in relativity we can go back to pre-relativistic theory by shifting the constant C towards zero (which takes us back into Newtonian absolute time), let us see what happens if we push the constant k towards zero: observation becomes gratuitous — it is the pre-cybernetical situation again — and action impossible. This brings us back to the theory of epiphenomenal perception. I am of the opinion that once we have really had time to digest these phenomena of the reciprocal conversion of information into negentropy, it will change our whole conception of the universe and, in particular, the laws of causality and finality respectively. There is no question but that these ideas are interacting with those of Bergson, as well as of Teilhard de Chardin.

Therefore I do not claim to be either for or against Teilhard de Chardin. What interests me is the problem itself; as regards Father Teilhard, whom I knew and liked as a man, I shall not adopt a partisan attitude. I think it is much more interesting to think about the actual problem, as Bergson did and as Teilhard did, and to develop it without any cult of personality.

Reverend P. Leroy I have no intention of joining in the purely scientific debate on either Father Teilhard' ideas or his work. What I want to do is to give you some idea of the kind of man he was: the man I knew and with whom I was privileged to live for several years in succession, which was something quite extraordinary for a man like him who never stopped more than six months in the same house, having the itch to travel all over the world.

I feel considerable emotion in speaking of this man who was my friend — because it is about my friend that I want to tell you. I have known him for a long time. I first met him in 1928 — he had just returned from China and I, as a newly graduated student, had been assigned to China by the superiors of my Order. It was obvious I ought to meet this man who had just come back from there. I have an unforgettable picture of our first meeting: it was in a cellar, an office in the basement of the Museum of Natural History, here in Paris, Place Valhubert. Teilhard was working there on the collections he had brought back from China. He did not know me and I did not know him. All we had in common was, first, the Order to which we belonged, and the work we were going to do together. When I greeted him, he got up from his chair, sat casually on the edge of the table and invited me to take his place, then for a whole hour, with his penetrating gaze fixed on me, he emitted a presence which I cannot define but which left a vivid imprint on my soul: I cannot recall these memories without deep emotion.

We remained friends. I saw him again in China in 1931, as he was preparing to leave for the Yellow Cruise. We met again in 1939 at Tien-Tsin, then in Peking. Throughout the eight long years of the Japanese and Sino-Japanese war we were interned in a small house in Peking and I lived with this man who combined in his person the exceptional qualities of both a man of science and a perfect gentleman, devoted because of his fundamental goodness to the cause of mankind. Father Teilhard was the friend of man and of all men.

You will realise that I cannot here go into all the details of his life, but I would like to tell you a few things which will perhaps help the debate along. Father Teilhard was always a man of exceptional faith; he had the courage faith confers, something like that Paul of Tarsus had, and already at the time he was studying philosophy and theology he was haunted by the problem of man and his destiny. From both the geological and palaeontological aspects he was obsessed by an idea: to uncover man's origins, not in order to demolish what had been constructed before but rather to find via the past the guiding thread of our action. Teilhard said, wrote and thought that the only thing which interested him was the destination. If, therefore, the bases on which his philosophy rested were the past and experience, this was with a view to finding the future and the ultimate purpose.

His scientific training, on top of the philosophical and theological training every priest receives when he prepares to take Holy Orders, gave him the ballast he needed to embark upon the problems which occupied his mind. I would say that Teilhard had a very positivist notion of the situation of man. But at the same time he had a positive mysticism about his vision of man. It was the combination of these two points of view which made him what he was.

Science, when it goes far enough, when it pursues reality into its last defences, when it is not just content to assemble documents but tries to understand them too, connects up with metaphysics. For Teilhard, science had reached a crucial point: he wanted to place himself by means of his learning in a position from which he could overlook the whole future of man. Teilhard was a whole man, a real man, a man who loved his fellow men. I could quote plenty of instances to show how he deprived himself utterly for the sake of others. But I will just tell you one small anecdote which comes to mind and which I do not think has been told before. It happened in Peking in 1939. He had just returned from America and, as he belonged to an American group which paid him a monthly allowance, a fairly large sum had accumulated in his absence. When he arrived in his office he found a sizeable check. Now, in his laboratory there was a woman whose husband had deserted her and who was alone in the world. Teilhard did not hesitate for one second, he handed the check — it was for over $ 4,000.00 — to the woman and promptly forgot all about the matter.

Teilhard was a man stripped to essentials. He was the prototype of a humanity which sees, lives, sympathizes, suffers and gives. The last time I saw him in December 1954 in the streets of New York — we were walking from his office to the little restaurant where he used to have lunch — he said something to me which has the profoundest resonances for all Christians: "I can tell you now that I live constantly in the presence of God."

This was the man who was also a great scholar with an international reputation in his speciality. He was a philosopher who had searched his conscience regarding the value of the answers science is able to give to the problems which preoccupy us; he was a theologian who made the effort to combine with the philosophical ideas which flowed from his fundamental experimental work the revelation which inspired his life. And last of all, Teilhard was a mystic. It would require a very long lecture to do justice to all his qualities. But perhaps from this brief and imperfect sketch you can imagine in more detail the various sides of his rich personality.

Helmut de Terra There is very little that I could add to the characterization — the very endearing characterization given by Father Leroy of Father Teilhard de Chardin. I have had the good fortune to have been associated with Teilhard on two occasions, one in India in 1935, and two years later in 1937, on an expedition in Burma and Java. During that time there were certain characteristics that stood out and stand out very clearly in my memory. One is that Teilhard de Chardin pursued his daily tasks, his very arduous tasks as an exploring palaeontologist, as a collector of fossils in those barren canyons of north-western India, with a deep joy, with great energy and with great scientific zest — a zest that was really so impressive to me. When on occasion I became tired after twelve hours' walking he said, "No, my dear friend, we must go on, because if *we* do not find another site who will find it for us?" Such statements were repeated many times.

Now, another very striking characteristic of his was that he was reluctant to engage his collaborators in any philosophical or theological discussions. Of course, he could not very well have done this with me because I was just a geologist, a prehistorian with whom he discussed the problems of the day — where to find certain remains of primates or of prehistoric men. But even at times when I came to ask him for guidance in moral conflicts, he showed great respect for the other person; he had what I think is so very rare in intellectual people, a real love for his fellow-men.

Now, it seems to me that Teilhard de Chardin has given his science a certain new dimension. As a student of earth history he foresaw the importance of biochemistry for elucidating the role played by the organization of matter on a molecular level — what he was wont to call la "pré-vie". I have been told recently by a former friend of his that among his unpublished letters there is one in which he stated that, if he were given a chance to start his career as a scientist all over again, he would start it as a biochemist. This is a very extraordinary statement to come from a palaeontologist, an Earth historian. It shows his great vision of what biochemistry might teach us in the way of what he called la "pré-vie" — the primary organization of matter.

As a palaeontologist and prehistorian, Teilhard de Chardin attempted to relate the deployment of evolutionary processes to man. His concept of "anthropogenèse" relates the emergence of a self-reflecting organism to the gradual evolution of consciousness by the increasing complexity and diversification of organic forms. He has, I think, given new meaning to palaeontology by lending priority to the human phenomenon, reminding us of the vital necessity of relating all science to man; and here is something we might call the convergence of thought in our time. A French philosopher, Georges Gusdorf, of the University of Strasbourg, who participated last year in a symposium arranged by the *Wenner-Gren Foundation for Anthropological Research* in Zurich, pointed out that in order to make science the servant of mankind, we must relate all science to man; in other words, we can no longer afford to get lost in specialization, we must come to a synthesis which can only be found in a new philosophic meaning of what science has done in relation to man.

In this respect, Teilhard's anthropological approach would seem to be akin, perhaps, to the relativism of modern physics. More important still, his experience of science transcends its traditional limits in that his concept of life and man lies in the realm of what he was wont to call the ultra-physical or metaphysical. From that arose his vision of a synthesis of religion and science.

The great question before us, it seems to me, is how to proceed further in an attempt to give philosophic significance to science, and whether we are, in our time, prepared to accept a transcendence of science. Sir Julian Huxley calls for an ideological, a humanist frame, while Teilhard would have us believe in a cosmological Christ.

I think it would be a wonderful thing if, under the auspices of Unesco, there could be set up some kind of a working committee of scientists who would find a common language in the framework of the ideas of humanity of Teilhard de Chardin.

Pierre Chouard I think what we are here to do, under this theme "knowledge of nature and man", is to try to understand Teilhard's contribution to a unitary concept. That is the meaning of this whole colloquium. After all that has been said already, I can be brief. I am, together with Mr. de Terra and Father Leroy, one of the three witnesses — though there are others present here — to a large part of Teilhard's life. What strikes me most of all forty years later — for it is forty years since I had the joy of knowing him — is how his expressions, now quite familiar to all, were a real revelation to a young science student just beginning his studies. At that time they were an unmatched innovation, capable of generating enthusiasm for uncovering something new;

now, of course, these expressions which I shall briefly recapitulate are so familiar as to seem almost banal. I am not going to tell you anything new; the point is that all this was contributed by Teilhard.

So what did we see? We saw in front of use, young science students of the Ecole Normale as it then was, someone who had at times been described as a great biologist. Now, he was not a biologist in the modern sense of the word, "molecular biology", which he predicted but of which he could not then have anything like the knowledge we have today. But he was unquestionably a great palaeontologist and geologist, a man whose interest one might expect to lie in the past; yet his whole attitude led us toward the future, that is toward the development of life. This was the first surprise. Then again, his vision embraced the whole of nature, the whole cosmos, where he assigned the leading role to life; it was this that made him a "biologist". The second surprise was that he not only made clear the entire development of palaeontology as expressed in evolution but he made its history definite and capable of supplying an intellectual model for the development of the entire cosmos, with rules which brought it to its end point, which is man. It was a revelation indeed to realize that the essence of this evolution was that "increasing complexity" which has been described to you, and that at the end point of this complexity appeared not only additional elements of "information" — which may become negentropy, as Mr. Costa de Beauregard was telling us — but also something quite "new": this is that at a certain level of complexity there appears self-awareness, the self-critical moral sense. And from this time on, along with the possibility of clearly and purposefully shaping the future, there also appears a model to be attained and hence a vital means of acting, also purposefully and consciously, upon evolution itself.

Here was the expression of an evolutionary continuity which can assume an almost explosive aspect when it reaches the level of awareness which man has reached and then goes on to seek new evolution. For evolution is continuing, but its principal actor is no longer the whole of nature moulding itself in some way, the model is now man, and man thus becomes the agent of evolution. Thus, from "alpha", the point of departure in matter, up to this whole luxuriance of the living animal and vegetable world and of man, there opens before us a prospect whose duration is perhaps unimportant, a prospect which is in process of becoming oriented towards the point "omega". Thus the whole world is comprised within a single movement of evolution, having a number of variants, true, but strictly one movement, strictly unitary.

Within this perspective, Teilhard started from the past to finish in the future; man is at the heart of this development and man charged with making the future is in a constant state of dynamism and hope. Hope is vital to happiness! Happiness is not only here and now; it is never in the past but is seen in its pure form only in hope, in what is being done, what will be done and in what we have a share in. The joy of conscious creation rests upon the freedom and responsibility to choose between values of which man is, at least, the norm. I think we have here one of the essential elements of Teilhard's optimism, which has sometimes been held against him; but Teilhard, who also saw the negative elements of the weakness of human nature, assigned to them only a subordinate place in his vision of the future. In this way Teilhard developed a new kind of humanism, having its roots in the old traditions and in the history of the past. It is also apparent how this humanism fitted in with his faith, so that there is no contradiction of any kind between his faith in what is revealed and his vision of the world. Yet at the same time his language can be understood by all men, whatever their religious or cultural affiliations. He thus brings us a system of unity, a system for linking all men, a unitary trend which is in the mainstream of the thought of this colloquium.

So this is my contribution, to tell you how "biology" can be the model for our future thinking. So now what has been called the "prospective", philosophy action to shape the future, can be seen to grow out of the perspective revealed by Teilhard.

Madeleine Barthélémy-Madaule What we have heard about Teilhard shows a touching combination of the scholar or thinker with the man. As a man, Teilhard was at the crossroads: he was a scholar, a thinker, a philosopher in the widest sense of the word (possibly a philosopher who cared not at all for the accepted meaning of the word philosopher); he was also a believer and no doubt a mystic, too. He wanted to unite all this — existentially — in himself. But the fact that he had this consuming hunger and thirst for unity would not in itself be a sufficient reason for our presence here today, at this colloquium with the theme "Science and Synthesis". Before he could justify all this, a man would have to have done much more than seek unity in himself, however intensely; a man would have had to make a viable objective synthesis which could be passed on to other men, not just a philosophy for the conduct of his own life.

Let us be honest, for to do homage involves looking the object straight in the eye — not that one need fear this with Teilhard. He is so controversial just because he is the prophet of synthesis. It is no dishonor — far from it — to be controversial. We have seen the urge to synthesis appearing among the scientists here. And yet, as a philosopher, I am acutely aware of the fact that there is at the moment an extraordinary harking back to analysis, a trend which Teilhard rightly said dominated the nineteenth century. Faced with the compartmentalism of science, Father Teilhard's reaction was to synthesize, to break down the partitions. The last book edited by Father de Lubac, *Blondel et Teilhard de Chardin*, includes a study in which Father de Lubac admirably describes Teilhard's reaction against this trend in today's disciplines. He even wrote Father de Lubac a letter saying that he would doubtless be thought reactionary because of his attempts at synthesis. Some people, of course, did hold this against him and we have to come to his defence. Teilhard saying his support for synthesis might seem reactionary, Teilhard telling a congress of palaeontologists he might seem reactionary because he defended orthogenesis (synthesizing vision) — this reminds me of the youthful Leibniz walking in the woods of Rosenthal and saying to himself that, while it might seem very modern to go along with Cartesian mechanism, one should perhaps leap over Descartes to link up with the substantial forms of the Middle Ages because this was the road to progress. Leibniz must have seemed reactionary, too. Teilhard was swimming against a very strong tide, for it is a fact that as the field of knowledge grows, man defends his position by making very detailed analyses. Just now the field of knowledge is growing faster and faster, till one grows giddy with it all, and man is horribly afraid to look beyond the end of his nose. He clings to his analyses because he is frightened of making premature syntheses. He is also frightened — perhaps with some justification — of getting his levels mixed, for every level has its epistemological statute and it does not do to muddle them.

So Teilhard must be challenged on these two points. Did he mix his levels? Did he really overstep the rigor of analysis in what might be called an irresponsible manner? If we had all day, we could take a theme by way of example and examine this example to see whether Father Teilhard, in making his synthesis, failed in the rigorous analytical approach and confused levels which do not mix. The example I think appropriate to "Science and Synthesis" is one that picks up the theme of this debate: knowledge of nature and man. Teilhard is sometimes, even quite frequently, reproached with having mixed the biological and historical levels, biology and anthropology, the natural and the human sciences. He is accused of shuffling ideas from one level to another, of having extrapolated biological analyses to apply them to man. So here we are in the thick of the battle for the spirit of synthesis. Mr. Gusdorf, quoted by Father de Lubac and a short while ago by Mr. de Terra, rightly said that epistemological frontiers have been fortified, have become watertight compartments and that the idea of breaking them down paralyses intelligence and imagination alike. Each discipline jealously guards its patrimony from incursions by foreigners. This results in a fragmentation of knowledge, weighing heaviliy in the tables of high culture. Father de Lubac also quoted some words — very topical because they were uttered by Professor Jacques Monod, a Nobel prizewinner, yet a man who shied away from a grand synthesis. Monod writes that new sciences nearly always grow from the invasion of one science by the ideas and methods of another; we could substitute "fertilization" for "invasion".

Nevertheless, in this area of the biological and the psychological, it is quite surprising to find how little Father Teilhard really mixed what he has been accused of mixing and how little extrapolation he did. Naturally, I cannot quote chapter and verse here, but if you will look at the texts where he writes about the biological nature of the social fact (which shocks many analysts), you will find that Teilhard covers himself. The phrase in which he relates the biological to the human always contains some saving expression like: "from the point of view from which we are looking at it" or "as regards the fundamentally mechanical aspects" and so on. So we see that in reality he highlights the features in man that relate him to the rest of life but without reducing the human to the purely biological. He just uses the biological to illustrate one aspect of the human. The very way in which he manipulates the two levels expresses the essence of what synthesis can mean in his hands. It is a delight to see, for example, in the paper on racism, how he puts it on trial — I think Unesco headquarters is a good place to quote this example; Teilhard faulted racism precisely because it uncritically extends the biological to the human without allowing for human specificity; then he carefully explains how error arises from confusing the two levels.

It is thus clear that while Teilhard seeks totality and a pattern of wholeness, he also gives the various levels their due. I think he left it to the philosophers, scholars and theologians to test the validity of the guidelines he sketched out; we ought to study on the basis of his own statements how he fits the levels together for each discipline individually.

In his book, *Problématique d'Evolution,* Mr. François Meyer has written penetratingly about the way he dovetails the laws for each level. There is no question of a mixed-up synthesis, on the conntrary, it is a structured one, and we should not forget that it comprises both an evolutionary and a dialectical synthesis. Teilhard also supplied the tools for this in his *Esquisse d'une dialectique de l'Esprit,* which is not read nearly enough; here he describes how thought should proceed from the best to the least-known fact, explaining its point of departure, the best-known, by reference back from the least-known, which is now better-known. After this, thought must take off from the now better-known least-known to enter a fresh stage which will repeat the two preceding ones, thus enabling further advances to be made. Thought is a spark which oscillates yet moves forward, progressing by checking back. This oscillating thought is universal, it is that of man who remains one while exploring all the watertight compartments, whose restless thought seeks totality by means of concepts it creates as it goes along; it can be generalized to suit all disciplines with no fear of confusion, hence it is safe to ignore the compartments. How can a universal language of synthesis be created via an evolutionary dialectic?

This, I think, is the reason why we are here. We arrive in this way at a totality which is not at all like that currently being sought by the structuralists, for example. They are looking to structure and analysis to provide totality. We know what this is; it is a totality of the linguistic type, though we can also call it psychoanalytical or Marxian-political. Each element in the structure refers to other elements of the structure; it is a state of perpetual reference to something else, and from these reciprocal references no sense of direction ever emerges! Now the sense of direction was what Father Teilhard was looking for. In the last analysis, what he seeks via his synthesis is the meaning of things; it was above all with regard to human phenomena that he needed to seek synthesis through the dialectic explained above. For the human phenomenon had been distorted by the scientists (I am sorry to have to say this, though they were certainly not the ones we see here today) for the very reason that they no longer took into account the whole of it. It had been distorted by the philosophers, too, because their view was equally unbalanced. While the scientists took an objective view of man in his physical (after all, we are physical, our bodies fall in space) and biological form (for we are biology), anthropological, too, (for we can still be studied objectively at the anthropological level) — while the scientists were busy fragmenting the human phenomenon, the philosophers were elevating him into something transcendental, almost royal. For them he was *the* subject, never regarded objectively, transcending the world under the banner of a philosophy of eternity!

But all this was partial and hence misguided. Father Teilhard called upon the subject to look at himself objectively, to look at and interrogate man-as-object; the oscillating spark of his dialectic passed from the ex-subject to the man-as-object, returning with the knowledge so gained to the subject, from man-as-object via the human and natural sciences, in order to ask fresh questions. The subject says to himself, "If this is what men are, if this is what *we* are, what are we to do in future?" We observe the play of this dialectic, man in turn seeing himself from outside, looking at himself in the mirror of objectivity, then suddenly turning back to himself, collecting his thoughts and anxiously asking himself, "Where I am going? What is the human condition? What is the meaning of it all?" To some this dialectic may appear naive but it is none the less true that it is essential to man, he cannot live without asking questions like this. But now men have to look in the mirror of anthropology and the natural sciences when they ask such questions. And so, Teilhard's synthesis, far from being old-fashioned, points to the future and traces there a whole program of work and trials.

We shall not be true to Father Teilhard unless we put his ideas to the test. Indeed, it may be that what the thinker in him was seeking was already known to the mystic; but he never let the mystic influence the thinker. So the mystic waited in prayer and fervency until thought in its endless development joined him upon the peak where he was waiting. Teilhard contained within himself both these men for, as Aragon rightly says, "Men are dual and will be till the end of time." Thus, Teilhard was a mystic who had founnd what he was looking for, but a thinker who was still seeking. So we shall never be true to his memory unless we, too, are true to his mystical certainty and his philosophical questionings, which are still unanswered.

François Meyer The philosopher sometimes feels himself to be at a disadvantage when speaking of Teilhard de Chardin whose thought lies at two poles, one scientific and the other theological, for the philosopher cannot claim to speak for either science or theology. Since this is so, I shall discuss some epistemological ideas, as sometimes when the philosopher takes an interest in the work of the scientist, he learns a great deal, not only about what the scientist can tell him objectively, but also about the scientist's attitude as an individual, that is as public and private thinker.

If we look — purely on the epistemological level — for the meaning of Teilhard's work and maybe his intention, too, we may conclude that he contributed in no mean fashion towards placing the study of evolution in its proper position. Indeed, the science of evolution is split between two approaches — the geneticist's and the palaeontologist's. Now, genetics has made some astounding advances and for this reason it is sometimes considered the only truly scientific view of the problem of evolution. We must climb this slope again, correct the perspective and take as our first premise, at least, that it is up to the palaeontologist to tell us what really happened in evolution.

Here Teilhard had the breadth of vision and the open-mindedness of the geologist. Due to his geological training, he thought — had to think — in huge norms and time spans. The geologist learns to think professionally in terms of *duration;* this saved him from what I shall call the *micro-causalities,* in which the geneticist too often immures himself, and he was able to grasp what we might call the phenomena of *macro-evolution,* for the phenomena of evolution require something like the order of magnitude of geological time before they become apparent. For the geneticist, used to investigation through laboratory experiments, such an order of magnitude seems an excessively remote approach, completely devoid of meaning in terms of genetic thought and knowledge. The lesson Teilhard taught us — and this proves his genius — was to think in terms of macro-evolution. It should not occasion too much surprise that it is possible to develop a true science of evolution which is not initiated at laboratory level and devoted to micro-causalities and micro-temporalities. It is a perfectly general tenet of positive epistemology that any science defines certain magnitudes of state by the level of observation it assumes. Thermodynamics clarified the distinction between micro and macroscopic levels; however, in one sense this distinction might be said to be inherent in all sciences, of any kind whatever. I would not defend this point of view here but it is pretty obvious when it comes to the science of evolution.

Evolution assumes its true dimension — which is cosmic — the moment one has the scientific "courage" to tackle the phenomenon of macro-evolution without thinking this is virtually a frivolous view, devoid of serious content, and resting on the excuse that the details are not to hand at the moment. Today we know that living matter can be traced back for at least two-and-a-half thousand-million years. Cosmologists now date the formation of the earth and the solar system — or, for some models of the expanding universe, the formation of the universe and the exploding of the primitive atom — at a time span of the order of ten to fifteen thousand-million years. Hence the duration of biological evolution is no longer trifling, as was still assumed in the nineteenth century, relative to the duration of cosmic evolution. It can be legitimately claimed that the evolution of life is of a cosmic order of magnitude, at any rate, regards its duration. This is accepted by science today.

A third point I wish to bring out is that the idea of evolution had long since ceased to be novel. Father Teilhard, his ideas and his work, met with no serious opposition from the Church. Any reservations which might have persisted, within the Church or elsewhere, concerning the evolution of species now seem to have vanished completely. Father Teilhard did not have to teach us that living things had evolved — this was known already. Wherein, then, lies the originality of his evolutionism, from a strictly scientific angle, compared with what I don't doubt is the version still adhered to by the great majority of evolutionists? It is this: Teilhard was sure evolution was going somewhere, that it had not only a direction, but a *goal*. It was directed towards an end where it would cease *qua* biological evolution. Of course, this is an assumption in Teilhard's perspective, a theological-type parousia; but, as a scientist, he could not omit this condition from his vision of evolution. This is probably the part most evolutionary scientists find hardest to swallow; they like to speak of "indefinite" evolutionary progress, undirected and open to "any" future possibility. This is the Bergsonian notion of unending progress, open-ended evolution. It is also the attitude adopted in dialectical materialism towards the universe and life, and this is the prevailing concept today. Dozens of reasons are given in support of this view, but the one I consider of philosophical interest is that to assume that biological evolution will stop short at man is to display a deplorable anthropocentricity. We have, for instance, Bertrand Russell asking what would happen if the history of evolution were written by an earthworm. To which we can only reply with dignity that the earthworm is not in fact writing the history of life and the universe. Despite protests evincing a type of false modesty which is itself a departure from scientific objectivity, man — whether he likes it or not — occupies a dominant position: he is the arrow, the spearhead of evolution. We could rephase this in philosophical terms and say that it is characteristic of man that he pursues science. Thus, anything we may say *about* science or as a *result* of science presupposes man's handiwork. And what is man? He is a being who possesses the peculiarity — unique in the universe — of "recreating" the universe by his knowledge. Man's meditations mysteriously reflect the universe. This is a very queer situation which has come about only once in the course of evolution — with man.

We have assembled here to talk about evolution, the universe, entropy and time, and we should have the metaphysical honesty to admit that this presupposes that the history of life cannot be constructed without placing man at the point where life has succeeded in recreating through knowledge this mysterious reflection of life and the universe. This is a far cry from the earthworm; it is absurd to speculate what sort of story the earthworm would have made of evolution. Science is man-made and it is this fact which locates man at the center of the perspective; this is not subjective — he did not put himself there — it has objective validity. The science of evolution begins with man. The philosopher must not allow the metaphysical pregnancy of the situation to escape him.

I shall now leave these rather philosophical considerations and turn to some more objective and verifiable points drawn from scientific observation. The science of evolution is changing. I am thinking of Dobzhansky and Schmalhausen, Huxley, too, whose study of the mechanisms of evolution led them to make certain most interesting discoveries, from both the epistemological and biological points of view. These mechanisms, previously classed under the broad — and, be it said, obscure and somewhat vague — heading of selection, are now themselves evolving from the Darwinian or neo-Darwinian mutation/selection systems. Schmalhausen stresses this most strongly. The actual conditions under which selection operates are now being studied, the way the mechanisms controlling evolution change in the course of it and determine its direction. They are becoming ever more efficient at ensuring qualitative and quantitative variation in selection which in turn brings about an *inevitable* acceleration as the mechanisms become more efficient. This acceleration is readily identified in a number of phenomena — Florkin's biochemical phenomena or those of cerebralization — which can be represented by rising curves showing a typical slope. This steep rise, indicative of violent acceleration, suggests that evolution is indeed approaching the point where it is accelerating or fast that some decisive event in the history of life must be due in a more or less distant but certainly not indefinite future.

One could demonstrate on the basis of these and many other examples the profound transformation now going on in the field of evolutionary mechanisms, thus proving that the idea of *progress* in evolution, of a typical and constant *acceleration,* of a goal towards which all this inclines, is no longer foreign to scientific objectivity.

The science of evolution seems to be breaking free from the prejudices of the antifinalists, by which it had been obsessed and which were merely a reaction to abstract finalism, and to be ready to embrace a more concrete vision of reality. And in this reality of all things, the history of life can now be seen to be an orientated phenomenon, accelerating towards its accomplishment.

One more word on an aspect of Teilhard's evolutionism which is not always appreciated but which places him among the great cosmologists in the philosophical sense of the word. This is the role assigned to matter in his scheme of things; no longer is it subject to a mindless inertia, it is the seat of a vigorous dynamism. Matter itself, with no outside help, no intervention to finalize it, no "little push" from God, by its own indwelling impulse and its tendency to become more complex, gradually effects and thereafter progressively accelerates the evolution of life. This vision is the reverse, for example, of Bergson's; for him, matter is the garbage of life rather than its mainstay or the expression of its own material dynamism, sometimes he even sees it as an obstacle in the path of life.

Today studies of pre- and proto-biotic states seem to be confirming in a way we do not yet quite understand a kind of ferment, which appears to be more than a mere stochastic phenomenon. So matter may contain within itself something which could be the key — if not of all manifestations of life — yet of the forward motion so typical of life. We find in the neo-materialists what we find in Teilhard himself — a vote of confidence in life.

Since we are gathered under the sign of synthesis, the ideal science would, of course, be one which effects a true unity of all the aspects involved — pure matter, life, man and thought, and finally the cosmos. Is such a synthesis possible? Mrs. Barthélémy-Madaule was somewhat sceptical, almost uneasy, about the possibility of total synthesis. I must admit I am of the same opinion: it is simply impossible for science to complete the synthesis and accomplish totality. If we scrutinize the march of events throughout human history, we can see that the totalities were never essentially scientific; they were always far more total than science can afford to be. These totalities are *cultures* and in history cultures represent the thousand-and-one ways in which man makes into totalities such things as technology, law, action, art, theology. He does this not just scientifically but with his whole being — his mind, body and soul, his past, present and future, his status in the world. You may argue about the duration of such totalities, but they are man's vocation.

Perhaps it is no bad thing to say out loud in this Unesco building devoted to universal culture that even today we hardly know what makes a culture, yet our evolution is accelerating so fast that its relation to cultures is essentially one of demolition. Technology, above all, is shaking cultures to their foundations, to the extent that they are tending to disappear. And it is no idle question — it is on everyone's lips — to ask whether culture is still possible, whether mankind is capable not only of making a quantitative totality but of rediscovering the meaning of totality in its full stature. This is a very serious matter and perhaps one that should not be separated from that future which Father Teilhard presented to us as finite and imminent. We may not have endless time in which to create our earthly paradise. It is not impossible, though naturally I do not presume to prophesy, that the acceleration of which we are now aware, in and around us, should increase to the stage where we no longer have time for syntheses and totalities, at least, not in the forms we knew in the past. Perhaps we shall have to invent radically new forms. Gaston

Berger used to say that the great problem of education in our time was not so much that of learning to adapt to new situations as of learning to adapt to constant change, in other words of learning the art of pure adaptation, the biological cumulation of all the long, slow adaptations along the road of the evolution of life.

Clearly man stands today in a particularly problematical position and I doubt whether even the relief expected from the social sciences, following upon the material and life sciences, will be able to help him. The social sciences are no more competent than the others to effect total synthesis. Perhaps I am being unduly metaphysical if I declare that the idea of totality is beyond the puny resources we can muster — the tactical maneuvers of filling in a breach here or cobbling together a partial unity there. Let us not underestimate the order of magnitude of the problems involved in the destiny of mankind. We cannot solve these vast problems in our speeded-up time by the mere exercise of goodwill. Teilhard has challenged us to rethink Bergson's appeal for "more soul" to enable humanity to accomplish its destiny.

Index

Speakers at the colloquium

Auger, Pierre: p. 141, 153, 156, 159, 163, 164, 166, 167, 172, 175, 176

Barthélémy-Madaule, Madeleine: p. 194
Broglie, Louis de: p. 78, 95

Chouard, Pierre: p. 192
Cocconi, Giuseppe: p. 117
Costa de Beauregard, Olivier: p. 101, 112, 135, 140, 186

Debeauvais, Michel: p. 161, 175
Destouches, Jean-Louis: p. 146
Dubarle, Rev. Dominique: p. 17, 145

Espagnat, Bernard d': p. 126, 127, 128, 142

Gonseth, Ferdinand: p. 3, 99

Heisenberg, Werner: p. 12, 125, 127, 128, 130, 131, 132, 134, 137, 139, 142, 144
Holton, Gerald: p. 45, 156, 168
Huxley, Sir Julian: p. 28

Kedrov, B. M.: p. 70, 151, 168
Kourganoff, Vladimir: p. 100, 120, 165

le Lionnais, François: p. 77, 98, 101, 103, 124, 134, 173
Leroy, Rev. Pierre: p. 166, 184, 189
Lichnerowicz, André: p. 88, 97, 109

Maheu, René: p. xi
Matveyev, Alexei: p. 93, 138, 141, 142
Mavridès, Stamatia: p. 118
Merleau-Ponty, Jacques: p. 114
Meyer, François: p. 198

Oppenheimer, J. Robert: p. 8

Piganiol, Pierre: p. 154, 164
Piveteau, Jean: p. 179
Poirier, René: p. 103

Russo, Rev. François: p. 84, 96, 169

Santillana, Giorgio de: p. 37

Terra, Helmut de: p. 191
Trautman, Andrzej: p. 124

Ullmo, Jean: p. 86, 98, 129, 140

Vigier, Jean-Pierre: p. 91, 97, 128, 131, 132, 133, 143

Persons referred to during the proceedings

Abraham, Henri: p. 58
Adler, Friedrich: p. 63
Aeschylus: p. 40
Alembert, Jean d': p. 45
Anaximander: p. 37, 38, 39
Andrade e Silva: p. 80
Appell, Paul: p. 157
Apollodorus: p. 43
Aragon, Louis: p. 197
Archimedes: p. xiv
Aristarchus: p. 38
Aristotle: p. 38, 42, 43, 44, 134, 141, 188
Aubigné, Agrippa d': p. 41
Auger, Pierre: p. 151, 152, 169, 172, 174

Baer: p. 102
Balmer, Johann Jakob: p. 128
Barthélémy-Madaule, Madeleine: p. 201
Bayes: p. 187, 188
Berger, Gaston: p. 202

Bergson, Henri: p. 114, 159, 179, 186, 189, 201, 202
Besso, Michele: p. 3, 4, 47, 48, 49, 52, 55, 59, 63, 68, 69
Biedenham: p. 132
Biot, Jean-Baptiste: p. 42
Bleuler, K.: p. 126
Bohr, Niels: p. 10, 11, 65, 80, 81, 82, 83, 85, 126, 129, 140, 142
Boltzmann, Ludwig: p. 63, 97, 187
Born, Max: p. 65, 66
Bose, Sir Jagadis Chunder: p. 9, 15, 77, 81, 126
Bourbaki, Nicolas: p. 124, 173, 174
Brahe, Tycho: p. 21
Bridgman, P. W.: p. 50, 69, 158
Brillouin, Léon: p. 114, 188
Broglie, Louis de: p. 9, 77, 82, 84, 91, 92, 93, 94, 97, 101, 144
Broglie, Maurice de: p. 91
Bruno, Giordano: p. 22
Bucherer, A. H.: p. 59
Buffon, Georges-Louis de: p. 180
Busch, Adolphe: p. 11

Carnap, Rudolf: p. 49
Carnot, Lazare: p. 45, 188
Cartan, Élie: p. 90
Casimir: p. 128
Cauchy, Augustin: p. 125
Christophe: p. 175
Cicero: p. 39
Cohen, Robert S.: p. 49, 63
Compton, Arthur H.: p. 79
Comte, Auguste: p. 97, 161
Cook, James: p. 43
Copenhagen, School of: p. 82, 96, 134, 137, 139, 140, 142, 143, 144
Copernicus: p. 21
Costa de Beauregard, Olivier: p. 105, 193
Curie, Marie: p. 159
Curie, Pierre: p. 157

Dante: p. 40
Darwin, Charles: p. 32
Debeauvais, Michel: p. 163, 174
Debye, Petrus J. W.: p. 79
Dechend, von: p. 44
Descartes, René: p. 9, 18, 39, 44, 195
Dingler, H.: p. 51, 56, 63, 66
Dirac, P. A. M.: p. 112, 128, 144
Dobzhansky, Theodosius: p. 200
Doppler, Christian: p. 5, 120, 135, 140
Dubarle, Rev. Dominique: p. 85
Duer: p. 17
Dürer, Albrecht: p. 44

Eddington, Sir Arthur Stanley: p. 60, 110
Ehrenfest, Paul: p. 53
Einstein, Hermann: p. 46
Élisabeth of Belgium: p. 11
Ellis, Henry Havelock: p. 79
Eötvös, Loránt: p. 59
Espagnat, Bernard d': p. 127
Euclid: p. xiv

Fermat, Pierre de: p. 80, 114, 188
Feynman, Richard P.: p. 101, 112, 129, 133
Fizeau: p. 120
Flato, Moshé: p. 128, 130
Florkin, Marcel: p. 200
Fock: p. 91
Frank, Phillip: p. 50, 51, 56, 58, 60, 67, 97
Frazer, Sir James: p. 41, 43
Fresnel, Augustin: p. 79
Freud, Sigmund: p. 11, 48
Friedman, Herbert: p. 110
Fuchs: p. 173
Fuller, Margaret: p. 28

Gabor: p. 188
Galileo: p. 18, 21, 22, 63, 64, 157, 172
Gandhi: p. 12
Gödel, Kurt: p. 108, 111, 115, 116, 119, 120, 125
Goldstone: p. 15, 16, 17, 126, 132
Gonseth, Ferdinand: p. 24
Grossmann, Marcel: p. 48, 50, 52
Guérin, J. M. F.: p. 42
Gundel: p. 42
Gupta, S. N.: p. 126
Gusdorf, Georges: p. 192, 195
Guye, Charles Eugène: p. 58

Hallwachs, Wilhelm: p. 79
Hamilton, Sir William Rowan: p. 21, 80
Harrison, Jane: p. 43
Hartner: p. 44
Hawkins, Gerald: p. 43
Heisenberg, Werner: p. 126, 127, 128, 129, 130, 140, 142, 143, 145
Heller, K. D.: p. 55
Helm, George: p. 45, 46
Henseling: p. 42
Herneck, F.: p. 46, 48, 51, 52, 56, 60

Herwart, J. G.: p. 66
Hertz, Heinrich: p. 79
Herz, J. T.: p. 46
Hilbert, David: p. 48, 125, 126, 127, 134
Hofmann, Banesh: p. 91
Holton, Gerald: p. 39, 163, 164
Homer: 42
Hönl, H.: p. 52
Hopf: p. 65, 80
Hoyle, Fred: p. 160
Hubble, Edwin: p. 120, 121
Hume, David: p. 45, 55, 85
Huxley, Aldous: p. 37
Huxley, Sir Julian: p. 192, 200
Huyghens, Christian: p. 109

Infeld, Leopold: p. 49, 91

Jacobi, Carl: p. 21, 80
Jeans, Sir James: p. 181
Jordan, Camille: p. 124
Joule, James: p. 112

Kamerlingh Onnes, Heike: p. 46
Kant, Immanuel: p. 7, 45, 63, 86, 87
Kaufmann, W.: p. 58, 59
Kedrov, B. M.: p. 153, 154, 155, 156, 170
Kelvin, Lord: p. 140
Kepler, Johannes: p. 21, 38, 66, 100, 170
Kirchhoff, Gustav: p. 63
Klein, Felix: p. 48
Klein, Martin J.: p. 83, 96
Kolmogorov, André N.: p. 169
Kourganoff, Vladimir: p. 166
Koyré, Alexandre: p. 22, 86
Kraft, V.: p. 49

Lagrange, Louis de: p. 21, 89, 91
Lamb, Willis E.: p. 112
Lampa, A.: p. 55
Lanczos, C.: p. 64
Landé, Alfred: p. 135, 136, 137, 140
Langevin, Paul: p. 81, 98, 140
Laplace, Pierre-Simon de: p. 21, 91, 143
Larmor, Sir Joseph: p. 47
La Rochefoucauld: p. 147
Lavanchy: p. 58
Lehmann: p. 125
Leibniz, Gottfried Wilhelm: p. 135, 195
le Lionnais, François: p. 175
Lemaître, Georges: p. 110
Lenin: p. 70
Le Roy, Édouard: p. 180
Leroy, Rev. Pierre: p. 191, 192
Levi-Civitâ, Tullio: p. 90
Lévi-Strauss, Claude: p. 44
Licent, Rev.: p. 185
Lichnerowicz, André: p. 92, 98, 100, 101, 108, 114, 115, 117, 120, 123
Lockyer, Sir Norman: p. 43
Lorentz, Hendrik Antoon: p. 5, 15, 17, 53, 54, 84, 86, 87, 89, 91, 111, 126, 131
Lubac, Rev. Henri de: p. 194, 195
Lycophron: p. 43
Lysimachides: p. 40

Mach, Ernst: p. 13, 45, 46, 47, 48, 50, 51, 52, 53, 54, 55, 56, 57, 59, 60, 61, 62, 63, 64, 65, 66, 67, 68, 69, 85, 91, 97, 117
Mach, Ludwig: p. 56
Magnus, Heinrich Gustav: p. 102
Matveyev, Alexei: p. 95, 97, 98, 101, 105, 142, 143, 144, 146
Mavridès, Stamatia: p. 120
Maxwell, James Clerk: p. 5, 9, 21, 52, 53, 64, 79, 90, 91, 97, 114, 135, 145, 153, 154, 188
Mayr: p. 32
Mehlberg, H.: p. 188
Mendel, Gregor: p. 167
Mendeleev, Dmitri Ivanovich: p. 153, 154
Mercereau, J.-E.: p. 101
Merleau-Ponty, Jacques: p. 105, 119, 120
Meyer, François: p. 196
Meyerson, Émile: p. 54
Michel, Louis: p. 128, 131
Michelson, Albert: p. 5, 54, 84, 95, 98
Millikan, Robert Andrews: p. 79
Minkowski, Hermann: p. 4, 49, 50, 53, 60, 61, 89
Mises, R. van: p. 49
Monod, Jacques: p. 166, 195
Morgan, Lloyd: p. 31
Morley, Edward Williams: p. 5

Needham, John T.: p. 44
Neumann, John von: p. 136, 146
Newton, Isaac: p. 21, 22, 39, 46, 52, 53, 55, 61, 63, 64, 87, 88, 89, 90, 100, 101, 104, 107, 123, 124, 125, 141, 144

Oppenheimer, J. Robert: p. 38, 69
Ostwald, Wilhelm: p. 45, 46
Ovid: p. 43

Panopolite, the: p. 43
Parmenides: p. 44
Pascal, Blaise: p. 22, 114, 181, 188
Paul of Tarsus: p. 180, 190

Peirce, Charles: p. 9
Petzoldt, J.: p. 48, 50, 56
Piganiol, Pierre: p. 156, 164
Pirani, Félix: p. 52
Planck, Max: p. 9, 51, 52, 54, 58, 64, 66, 67, 69, 78, 79, 80
Plato: p. 44
Pliny: p. 42
Poincaré, Henri: p. 45, 54, 86, 91, 102, 111, 128, 130, 131, 132, 135
Poirier, René: p. 103, 110, 111, 116, 117, 119, 120, 123
Polemon: p. 40
Popper, K.: p. 135
Porphyrius: p. 43
Prandtl, Ludwig: p. 102
Proclus: p. 43
Ptolemy: p. 21, 43
Pythagoras: p. 173

Rayleigh, Lord: p. 98
Reichenbach, H.: p. 55, 112
Ricci, G.: p. 90
Riemann, Bernhard: p. 9, 50, 65
Roosevelt, Franklin D.: p. 12
Rosenthal-Schneider, Ilse: p. 60
Russell, Lord: p. 12, 68, 199
Russo, Rev. François: p. 86, 88, 91, 172, 173, 174
Rutherford, Lord: p. 11, 112, 150, 166
Rydberg, Johannes Robert: p. 126, 128

Schardin, H.: p. 55
Schilpp, Paul A.: p. 49, 50, 58, 68
Schlegel, G.: p. 42
Schlick, Moritz: p. 62, 67
Schmalhausen: p. 200
Schneider, Marius: p. 44
Schrödinger, Erwin: p. 81, 82, 90, 94, 125
Schwartz, Laurent: p. 174
Schwinger, Julian S.: p. 101, 112
Selig, Carl: p. 47, 59
Serre, Jean-Pierre: p. 174
Shankland, R. S.: p. 54
Shelley, Percy Bysshe: p. 40
Simpson: p. 32
Smoluchovski: p. 92
Snow, Lord: p. 160
Sommerfeld, Arnold: p. 17, 80, 112
Sophocles: p. 40
Spinoza, Baruch: p. 11, 23
Stallo, J. B.: p. 45
Sturm, Charles: p. 173
Szent-Györgyi, Albert: p. 34

Taton, René: p. 152
Terra, Helmut de: p. 192, 195
Thom, René: p. 174
Thirring: p. 17
Thomson, J. J.: p. 150
Tomonaga, Sin-itiro: p. 101, 112

Ullmo, Jean: p. 89, 95

Valéry, Paul: p. 148
Van der Waals, Johannes: p. 187
Van der Waerden: p. 44
Vigier, Jean-Pierre: p. 83, 94, 98, 129, 130, 145, 146
Vinci, Leonardo da: p. 38

Wegener: p. 184
Waddington, Conrad Hal: p. 31
Weinberg, C. B.: p. 61, 63
Weiner, Armin: p. 54
Weiss, Pierre: p. 99
Weiszäcker, C. F. von: p. 17
Werner, Alfred: p. 44
Wightman: p. 125

Yamamoto: p. 17
Yamasaki: p. 17

MIX
Papier aus verantwortungsvollen Quellen
Paper from responsible sources
FSC® C105338

If you have any concerns about our products,
you can contact us on
ProductSafety@springernature.com

In case Publisher is established outside the EU,
the EU authorized representative is:
Springer Nature Customer Service Center GmbH
Europaplatz 3, 69115 Heidelberg, Germany

Printed by Libri Plureos GmbH
in Hamburg, Germany